Carl Zeiss

A Biography   1816–1888

Published by the ZEISS Archives
Authors:
Stephan Paetrow and Wolfgang Wimmer

2016
BÖHLAU VERLAG KÖLN WEIMAR WIEN

Legal Information

Carl Zeiss
A Biography 1816–1888

Published by the ZEISS Archives

Bibliographical information from the German National Library:
The German National Library has a record of this publication in the
German Nationalbibliographie; detailed bibliographic information
is available online at http://portal.dnb.de.

Front cover (from left):
Carl Zeiss aged 34/35; photo by Carl Schenk.
Bench for mounting lenses from second half of 19th century.
Microscope stand I produced by Carl Zeiss in 1878.
The last page of the contract between Carl Zeiss and his son Roderich, dated August 1883.
Carl Zeiss circa 1870.

Back cover (from left):
Carl Zeiss's house and administrative building on Littergässchen in 1890.
Jena circa 1845, engraved by H. von Herzer.
The Zeiss men's choir in 1869.
Carl Zeiss at the beginning of the 1880s.

© 2016 by Böhlau Verlag GmbH & Cie, Cologne Weimar Vienna
Ursulaplatz 1, 50668 Cologne, Germany www.boehlau-verlag.com

Authors: Stephan Paetrow (timefab), Wolfgang Wimmer (Carl Zeiss AG)

Editors: Michael Kaschke (Carl Zeiss AG), Gudrun Vogel (Carl Zeiss AG),
Timo Mappes (Carl Zeiss Vision International GmbH), Kathrin Siebert (descendant of Carl Zeiss), Tim
Sander (timefab), Dieter Brocksch (Carl Zeiss AG), Marte Schwabe (Carl Zeiss AG)
Proofreading: Constanze Lehmann
Cover design: Bernd Adam, Jena
Layout and graphic design: Bernd Adam, Jena
English Translation: Clive Poole, James Humphreys, Marina Stephanou, Charles Taggart
Printing and binding: Finidr, Cesky Tesin
Printed on chlorine- and acid-free paper
Printed in the EU

ISBN 978-3-412-50423-6

Contents

Foreword

Traditionally, round-figure birthdays have always been occasions for honoring important people, not only by organizing special events, but also by publishing biographies. The 100th birthday of Carl Zeiss occurred during the First World War, his 125th birthday during the Second World War. Only individual newspaper and magazine articles appeared on both occasions. This made the 150th birthday of Carl Zeiss in 1966 even more important. The company had been partitioned as a consequence of the Second World War, and each site published its own book: Paul Esche, the former head of the company archives at Carl Zeiss Jena in East Germany, wrote a biography which appeared in two editions. In the town of Oberkochen in West Germany, it was the curator of the Optical Museum, Horst Alexander Willam, who rose to this challenge. A battle was waged during the Cold War over the sole right to depict and interpret the life of Carl Zeiss. But two members of the extensive family association, Erich Zeiss and Friedrich Zeis, also

prepared their own biography. They occasionally criticized the 'Abbe bias' of many historical publications which had originated within the company.

The first comprehensive biography of Ernst Abbe, the long-time partner of Carl Zeiss, appeared in 1918, i.e. more than 10 years after Abbe's death. Other, less voluminous publications followed. It was largely Abbe's academic colleagues and students who felt obligated to perpetuate his legacy. This was neither premature nor particularly comprehensive in comparison to other important entrepreneurs and academics. By way of contrast: it was first and foremost Ernst Abbe who reported on Carl Zeiss. His retrospective look at the company's first 50 years has remained an invaluable source both for the early history of the company and for the person Carl Zeiss. There are short reports from several contemporaries dating from the 1920s on the conditions prevailing in the company's early history. These appeared in the employee magazine, which had been recently introduced.

Carl Zeiss's most significant creation – the company named after him – speaks for itself. The company's success is rooted in its ability to still defend its market position more than 125 years after the death of its founder. Ernst Abbe continued the transition to a science-based company and further accelerated the enormous increase in the company's size and reach that had begun in the 1880s. It was also Abbe who established the Carl Zeiss Foundation, the embodiment of his entrepreneurial philosophy. This was intriguing for the intellectual public and a seminal moment in the history of German industry. Thus Abbe was more often in the spotlight than Zeiss.

The number of written documents left to us by Carl Zeiss is surprisingly small. The reason: until Zeiss's death in 1888, there was relatively little bureaucracy at the company. Meetings did not require written minutes and employees received their instructions orally, not in writing. The limited amount of material made it difficult to write an original biography. Therefore we need to make these few company records more readily accessible. Within the scope of a project sponsored by the Carl Zeiss Foundation, different sources are currently being digitized and will soon be available on the internet.

For the life of Carl Zeiss, three sources have provided new findings, some of which have made their way into this book:

- Carl Zeiss kept a 'Manual' – a ledger – where he documented the income and expenditure of his new company and his family from 1848 through 1863. This is an indispensable historical source for social and economic history. However, the handwriting in the ledger is difficult to decipher. At the moment only rudimentary evaluations are available because the amount of work required for a detailed survey is monumental.
- Evaluating the lists of employees has shown how and when the company's workforce grew, where the employees came from, which jobs were in particularly high demand and at what times the staff turnover was surprisingly high.

- The microscope ledgers contain detailed documentation on which microscopes were produced, to whom they were delivered and with which features. By focusing on when and where these instruments were delivered, we obtained reliable data covering a longer period for the first time.

Sometimes intensive research is also necessary to prove that an event did not take place. This is something that every archivist knows. People praise your efforts when you find something quickly, but when you search in vain, people are skeptical and think: did my colleague actually bother to look? The latter was the case when searching for the supposed rejection by the local authorities of Carl Zeiss's application to settle in Weimar to work as a precision mechanic in 1845. There was no evidence to substantiate this event in any archive. Thus it appears quite probable that the story is a legend.

The intention of this publication is to make Carl Zeiss's biography accessible to an interested public by making it very readable and visually appealing. The projects sketched out above show that this book is intended to be a starting point rather than an end point. I hope every reader enjoys this biography and discovers something new.

Wolfgang Wimmer
Head of the ZEISS Archives

Chapter 1

From Whence He Came:
A Foray into the Origins of
Carl Zeiss

《 Page 7: The parents of Carl Zeiss: Johanne Antoinette Friederike Zeiss, née Schmith (1786–1856) and Johann Gottfried August Zeiss (1785–1849).
Painting owned by great-granddaughter Dr. Anneliese Seeliger-Zeiss.

Weimar circa 1800. Copperplate print by Georg Melchior Kraus.

A Craftsman with Ambitions:
Family and Roots (1816–1834)

Famous for Goethe, Schiller, Herder and Wieland, the four most important German poets of their times, and for the Court of the Muses set up by the erudite Duchess Anna Amalia, Weimar was deemed the cradle of national culture in the early 19th century. To this very day – and in rare harmony – marketing blurbs aimed at inquisitive tourists and literary supplements targeted at a more scholarly audience sing the praises of Weimar as the city of poets and philosophers. However, more recent research has shown that although this image is not totally wrong, it is very incomplete. In actual fact Weimar was more a city of craftsmen and, although the importance of trade and commerce was already growing, the city chiefly owed its economic prowess to them. In 1815 the workshops and residences of master craftsmen with their families, fellows and apprentices accounted for almost 40 percent of land ownership within the city. The most affluent customers came from the ducal family and the appertaining court household, including the upper echelons of the civil service. Luxury goods were in great demand, and this was reflected by the broad spectrum of skilled crafts.

Zeiss's parents, August and Friederike, came from Buttstädt in eastern Germany. Image from 1650 (Matthäus Merian).

Parental home

One of the master craftsmen who focused on the needs of the upper class was the "horn turner" and subsequent "art turner" August Zeiss (1785–1849). He had inherited the craft from his father who had also been a turner in the small town of Rastenberg and later in the slightly larger Buttstädt about 20 kilometers from Weimar. Despite the interest in science

passed down to him by his father, August Zeiss had very simple schooling. With the exception of his six months of journeymanship which took him all the way to Dresden, his father constantly encouraged him to work at the lathe. In view of his more lower middle class background, August Zeiss's wedding in 1808 must be considered an ascent on the social ladder. His marriage to Friederike Schmith (1786–1856), the daughter of the Buttstädt town magistrate and ducal court advocate, was definitely a beneficial match. As both August and his older brother remained loyal to the craft of turning, things gradually became too cramped for the younger man in Buttstädt. Together with his wife, he moved to Weimar in the same year as their marriage, where he then received citizenship at the age of 23. August Zeiss switched his attention to producing and trading smoking accessories as the following advertisement in the *Weimarisches Wochenblatt* on 4 March 1809 demonstrates:

"I herewith inform the most honored public that I have moved into the house belonging to Frau von Lincker on the Töpfenmarkt. I manufacture all sorts of genuine pipes and other fine wood-turning work. I also offer Meissen porcelain pipe bowls and the like at very cheap prices."[1]

Zeiss appears to have been a good businessman. As early as August 1812 he had saved enough money to purchase at auction a three-story house on Breite Gasse no. 53 (now Marktstrasse 13) for 1,540 thalers. The building was located right in the center of

Market in Weimar before 1837, engraved by Carl August Schwerdgeburth.

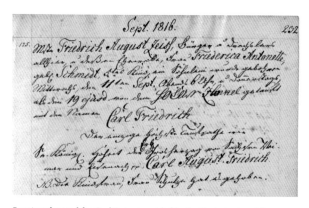

Baptismal record for Carl Zeiss, provided by the Protestant parish in Weimar.

Cover of the book "Die Drehkunst in ihrem ganzen Umfange", edited by August Zeiss, 1839.

Weimar, about half-way between the Theaterplatz and Marktplatz. The workshop was downstairs; Zeiss and his family lived on the first floor. He rented out the upper stories to the master clockmaker Eberhardt. It was in this building that the couple's fifth child was born on 11 September 1816. The son was baptized Carl Friedrich, after his only godparent, the heir to the throne of Saxe-Weimar-Eisenach. Carl Zeiss would eventually have a total of eleven siblings, six of whom died in their infancy.

A prince as his godfather

Crown Prince Carl Friedrich (1783–1853) played an interesting role. Having persons of elevated social status as godparents was nothing unusual during this time. Most citizens strove to establish beneficial connections for their children as part of the baptism procedure in order to promote their offspring's subsequent social advancement. Nevertheless, the fact that a member of the highest political class was prepared to play the role of godparent to a lowly citizen was a rare occurrence in Weimar society at that time. This was all the more surprising as August Zeiss did not yet enjoy the status of a court craftsman in the year of his son's birth – he was not appointed as a "court wood turner" until 1829. The fact that the future Grand Duke nevertheless became godfather to Carl Zeiss can be attributed to the long personal relationship with the latter's father. August Zeiss had been instructing Carl Friedrich of Saxe-Weimar-Eisenach in the art of turning since 1810. This was evidently one of the prince's hobbies which also included painting and wax embossing, i.e. the production of reliefs or sculptures from wax.[2] These private lessons gave August Zeiss direct access to the prince which he then used to secure an influential godfather for his son. However, the literature contains no evidence indicating that Carl Zeiss took advantage of his relationship with the Grand Duke's family during his subsequent career.

At this point it would seem appropriate to interrupt the account of Carl Zeiss's childhood and youth to make two source-critical comments: first, it is now known that Carl Zeiss was not born at the house on Kaufstrasse 1 directly on Weimar's Marktplatz. The commemorative plaque attached to this building for several decades was in fact on the wrong house. Nevertheless, the building at Kaufstrasse 1 had an important function because Carl Zeiss's father purchased it in 1818, and the family moved there when Carl Zeiss was two years of age.

The second aspect concerns the writing of the name which was probably derived from the German verb *zeiseln* (=to hurry, to be busy). This book contains

Portrait of the heir to the throne Carl Friedrich, after 1853; lithography print by Friedrich Martersteig.

The market square in Weimar (1850). Carl Zeiss grew up in the house second to the right of the town hall.

Change abounds

Let us return to the Weimar of 1816. Carl Zeiss was born into an era that saw the dramatic transition from a class-based, hierarchical society to a civil one. The nobility, clergy and even professors still lived in their own spheres and enjoyed their own privileges and jurisdiction. They kept their distance from the 'lowly' urban bourgeoisie, although there were occasional points of contact such as the Crown Prince's role as a godparent to Carl Zeiss. For some years, however, the winds of change had been wafting through Thuringian Weimar, and cracks were appearing in the city's rigid, traditional order. Education and personal achievement, and no longer birth and inherited estate, now became the factors determining who was appointed to important offices. The tight fabric of class barriers, guilds and family traditions was gradually unraveling, making room for the individual. This development began in all German states during the transition from the 18th to the 19th century, but it was at its most pronounced in Thuringia.

The marriage of Crown Prince Carl Friedrich to the Czar's daughter Maria Pavlovna meant that Saxe-Weimar-Eisenach now fell under the auspices of Russia. This connection was of great benefit when, after Napoleon's defeat, Europe received a new order at the Congress of Vienna in 1815. The Ernestine dukes in Weimar were granted a significant enlargement of their territories, including large parts of the State of Erfurt which formerly belonged to the archdiocese of Mainz. However, even though Saxe-Weimar-Eisenach

solely the spelling 'Carl Zeiss,' although our protagonist himself and his acquaintances also used other forms ('Zeiß,' 'Zeihs,' 'Zeis' and even 'Zeyesz'). His first name also varied between 'Karl' and 'Carl.' These variations were acceptable at the time in question. There was no such thing as long-term identity cards or compulsory registration. In most printed publications the name was already written with a double 's' during Carl Zeiss's lifetime instead of the letter 'ß' which is unique to the German language. After the name Zeiss became an international trademark due to the success of the optical workshop, Roderich Zeiss and Ernst Abbe agreed on the standardized spelling 'Carl Zeiss' in 1885 – probably also because the version with 'C' and 'ss' was more practical for international business correspondence. His descendants also used this spelling for their family name.

The church and the grammar school in Weimar, circa 1840.

now called itself a Grand Duchy, it remained a small state with limited resources. Here, the social distance between the political establishment and craftsmen and tradesmen was not as pronounced as, for example, in Prussia, Austria or Bavaria. The Grand Duke and his government were therefore prepared to improve the conditions required for the social ascent of lowly citizens – by promoting educational facilities and by reforming the guild charter in 1821, for instance. It was during this time of economic upheaval and cautious liberalization that Carl Zeiss spent his childhood.

Little is known about the early years of his life. What is certain, however, is that August Zeiss provided his sons with an extensive school education. Like his older brothers Eduard (1809–1877) and Gustav (1811–1875), Carl Zeiss attended grammar school. Next to nothing is known about his younger sisters Pauline (1818–1900), Hulda (1821–1888) and Emilie (1828–1875). As befitted the patriarchal ideology of the time, the three girls probably enjoyed only a modest education that prepared them for their roles as wives and mothers. No documents indicating how the sisters ultimately spent their lives are available. In relation to Carl Zeiss, the grammar school records show that he was admitted directly to the eighth grade on 13 September 1827. This would suggest that Zeiss had already received instruction elsewhere – either with one of the numerous private teachers or at the Weimar Bürgerschule (Citizens' School) that

The hunting lodge on Marienstrasse in Weimar, which also housed the Trade School; photograph taken by Louis Held circa 1900.

Carl Ludwig Albrecht Kunze circa 1855, photograph taken by Julius Schnauss.

was set up in 1825. On 29 March 1832, Zeiss left the grammar school with a qualification verifying that he had finished his penultimate year. This entitled him to study technical subjects at the university.

A new type of school education

In answer to the question as to why Carl Zeiss, unlike his brothers, did not fully graduate from grammar school but left in his penultimate year, the relevant literature points to the boy's health. Evidently Carl Zeiss suffered from a hernia and, in his father's opinion, should not conduct any undertaking that required long periods of sedentary activity. According to his father, this meant that he was predestined for a career in the technical crafts. This story is often mentioned in the records – despite its medical implausibility. Perhaps this was more attributable to the son's per-

sonal proclivity and to the family tradition of ensuring that at least one of the sons should remain a craftsman. As Carl Zeiss himself later reported, he attended what was known as the Trade School in Weimar that emerged from the existing Drawing School following an initiative launched by Johann Wolfgang von Goethe in October 1829. Held on Sundays, the Trade School was specially geared to the needs of complex crafts, with technical drawing as one of the focal points of its curriculum. One of Zeiss's teachers was a man with practical experience, the Weimar building inspector Carl Friedrich Christian Steiner (1774–1840) who was the architect behind the construction of the local Hoftheater (predecessor of the Deutsches Nationaltheater) and the tower of the Anna-Amalia Library. Another pedagogue who probably played a key role in Zeiss's education in Weimar was the North German mathematician Carl Ludwig Albrecht Kunze (1805–1890), who taught at the grammar school from 1828 and later at the Trade School as well. From 1830, Kunze set up a physics cabinet for the grammar school, gave experimental lectures to the public using mechanical instruments, wrote several textbooks and created geometry games with didactic content. Kunze's teacher colleague Ernst Christian Wilhelm Weber (1796–1865) attributed a pivotal role to Kunze in the modernization of the subjects of relevance to Zeiss:

"When you first assumed your position as a teacher, instruction in mathematics at our school was very meager as it barely addressed even the rudiments of geometry. That the teaching of this subject has now been broadened to meet today's needs is solely your personal achievement."[3]

Carl Zeiss was part of the first generation of students to enjoy an improved education which was more sharply focused on the requirements of technical occupations. His schooldays in Weimar may have been of key importance in the graduate's subsequent decision to devote his energies to the fields of mechanics and optics. While the young Zeiss was still acquiring the knowledge and skills he needed to pursue a career as a qualified craftsman, his father succeeded in further cementing his status in society. August Zeiss numbered among the founding members of the Weimar Trade Association, of which he was also the first chairman from 1834–1835. However, Carl Zeiss left his home city at Easter 1834 and chose a nearby destination to pursue his further career: Jena, a good 20 kilometers east of Weimar and the seat of the state university.

Cover of a textbook by Carl Ludwig Albrecht Kunze, 1866.

1   *Weimarisches Wochenblatt* no. 18, 4 March 1809, p. 78.

2   We owe this information and further helpful comments on August Zeiss to Dr. Ulrike Müller-Harang of the Klassik Stiftung Weimar.

3   Quoted in: Kerrin Klinger: *Zwischen Gelehrtenwissen und handwerklicher Praxis. Zum mathematischen Unterricht in Weimar um 1800.* Paderborn, 2014, p. 115.

Questions on Her Great-Great-Grandfather:
An Interview with Dr. Kathrin Siebert

How did you discover that you were related to the founder of ZEISS? Can you remember the exact moment you found out?

Not exactly. My historical knowledge of Carl Zeiss was something that grew over time. Our family did of course own a great many ZEISS products. We had one of the early microscopes, several pairs of binoculars and a whole range of camera lenses – our family had a real passion for photography. As an undergraduate and doctoral student of microbiology and biochemistry, I spent a lot of time using electron microscopes made by ZEISS. And when I began working in the pharmaceutical industry, I became acquainted with the medical technology produced by the two ZEISS companies in East and West Germany.

I slowly came to realize just how comprehensive the product range was.

I spent some 30 years of my career traveling the world. As someone in charge of quality assurance and quality management, I met countless people all over the world working in industry and science. Everywhere I went the name ZEISS was held in high esteem. Perhaps that's why the mention of my family history was often immediately met with such great interest. People's reactions were often especially positive in the US. Americans are particularly interested in finding out about your heritage. On numerous occasions, I also discovered that the ZEISS brand is very highly regarded in Asia.

So discovering that Carl Zeiss was your ancestor opened doors for you?

You can rest assured that, at work, my heritage was wholly irrelevant. My ancestry was, however, quite helpful when I wanted to visit ZEISS sites abroad out of personal interest. Even before the end of the Cold War, I felt extremely welcome no matter if I was touring a plant belonging to the East German combine or was at the company based in Oberkochen. I can still remember visiting a subsidiary of Carl Zeiss Jena in Wellington, New Zealand, before 1989 – that was certainly exciting given the political climate of the time.

Regular get-togethers are still organized for ZEISS representatives and the founder's family. This is something we all value very much and it is proof that Carl Zeiss is still an iconic figure today.

Let's travel some 125 years back in time. In May 1891 – approximately two years after Carl Zeiss passed away – the then family-run company was transformed into a foundation company under the direction of Ernst Abbe. Looking back, how is this move perceived by the family association?

The other descendants and I would certainly be interested to know whether ZEISS would have been as successful as it is today had it remained a privately owned company. As you can imagine, the topic has occasionally been touched upon at family gatherings. That said, you cannot lose sight of the historical context that prompted the decision to establish the Carl Zeiss Foundation: By the time he died in 1888, Carl

Zeiss had fathered four children. Roderich, born in 1850, joined the company in 1876 at his father's request. It is no secret that he and Ernst Abbe had very different ideas about the company's strategic focus. At the time, Roderich had no children and so it was not clear who would take over from him. Carl Zeiss had three children from his second marriage and they were not involved with the company at all: Otto was a physician, Hedwig married the principal of a grammar school and Sidonie became the wife of a doctor. In other words, there was no viable alternative other than to establish a foundation. At least not when you have two long-term goals, like Ernst Abbe did at the time: social security for the company's employees and the promotion of scientific research at the University of Jena. Under the circumstances, it was obvious to Abbe that the best way to achieve these goals was to depersonalize the ownership structure. This is what he did, and I can fully understand his logic.

Let's get back to Carl Zeiss: What sort of person was he? How does the family remember him?

His second wife, Ottilie Trinkler, was a tenth-generation descendant of Martin Luther. So it's eminently plausible to ascribe typically "Protestant" values to the company founder, such as a sense of duty, perseverance, the ability to see things through to the end, and precision. I would certainly not describe ZEISS as a company founded on Lutheran values, but I can believe that Carl Zeiss, an entrepreneur in precision mechanics and optics, possessed such traits. For me, Carl Zeiss's pursuit of quality is something immensely inspiring. Even so, I still see the workmanship associated with ZEISS differently than an ordinary customer

would. I have also experienced this through my work in quality management.

Carl Zeiss is purported to have taken a hammer to finished microscopes that did not meet his expectations – in front of his own assistants no less ...

This anecdote – if indeed there is any truth to it – is a fine example of how far the leadership style at ZEISS has come since the mid-19th century. There's not a single employee at the company today who would put up with such an outburst. Not only have times changed but the average age of people entering the workforce is now much higher. From a modern perspective the first apprentices to join the newly formed company were not much more than children themselves. At the time, it was perfectly natural for parents to accord the apprentice's employer certain parental responsibilities. Carl Zeiss was head of the company in precisely the way that was expected of him and acted just like any other entrepreneur of the time. We now know that patriarchy is not the best

recipe for success. This applies as much to the business world as it does to our family.

There is also historical evidence to suggest that Carl Zeiss mellowed with age. One report states that after retreating almost fully from the operational side of the business – and at the very latest following his first stroke at the end of 1885 – he discovered a passion for growing roses. My grandfather was born in 1886 and still had fond early childhood memories of Carl Zeiss. The little ones spent a lot of time at the home of the elderly Zeiss, who was truly devoted to them. This proves that Carl Zeiss cannot be summed up with a handful of adjectives.

That sounds like a warning to the biographer to choose his words wisely ...

We simply have to accept that there's not always a common theme. The life of Carl Zeiss was also colored by his inner struggles. This may also be one reason why it's worth retelling the life story of a famous man. We are making new demands of the historical material inspired by modern attitudes and getting answers that were perhaps not at all relevant for Zeiss biographers in the 1960s. For example, I think it would be worthwhile to find out how much of a say Carl Zeiss had in determining his successor. I know a lot of family-run companies in Germany that find it difficult to select an ownership structure that will serve them well in the future. This is something Carl Zeiss achieved early on and it could help inspire people today. I would also like to find out more about Carl Zeiss – the man behind the company. I'd be interested to know the extent to which this historical tradition can support our family history when we look at it critically and whether there are any aspects of Carl Zeiss's life that we have yet to explore.

....

Dr. Kathrin Siebert (*1954 in Hamburg) is the great-great-granddaughter of Carl Zeiss. She studied microbiology in Kiel and Miami and graduated with a PhD in natural science from Göttingen University. She worked in the R&D departments of various pharmaceutical companies and was also responsible for quality management. Today, she provides consultation services for companies on the production of pharmaceuticals and medical products.

Chapter 2

A Pioneering Spirit Is Born:
Apprenticeship and
Company Foundation

«« Page 21: Carl Zeiss at the age of 34/35, photograph taken by Carl Schenk.

Jena circa 1845, engraved by H. v. Herzer.

From a Small Town out into the World and Back:
Apprenticeship under Friedrich Körner and Journeyman Years (1834–1845)

In Weimar, trade and commerce were dominated by the Grand Ducal Court while in Jena it was the university that determined the weals and woes of the populace. The *Cives academici*, i.e. students, professors and other persons associated with the university, all of whom had their own jurisdiction, constituted quite a formidable economic power despite accounting for less than one fifth of the city's population. While the production of luxury goods flourished in Weimar, Jena had at its disposal a considerable amount of skilled tradesmen, such as coppersmiths, watchmakers and mechanics, who lived at least in part on work linked to science. Towards the end of the 18th century, however, a rather gloomy image of the city prevailed, as described in the following travel account penned by Friedrich Nicolai:

"Industry is rather scarce here; people expect to find sustenance partly from an inconsiderable amount of farmland, a small number of horseradish plants

Gustav Zeiss (1811–1875), the second-eldest brother of Carl Zeiss.

and a little livestock, but chiefly from the efforts undertaken by the students who are expected to bring them food just as the ravens did Elijah." [1]

After a short-lived intellectual upsurge in around 1800 associated with such names as Friedrich Schiller, Georg Wilhelm Friedrich Hegel, Friedrich Schelling, Johann Gottlieb Fichte and Friedrich Schlegel, the university experienced some dark times. Thereafter, the number of students only began to rise gradually. The family of the court-appointed art wood turner August Zeiss played a small part in this, with all three of his sons attending the eastern German university

Friedrich Körner, Court Mechanic.

in succession, from 1828 to 1838. Carl Zeiss went to Jena in early 1834 to begin an apprenticeship under the court mechanic Friedrich Körner (1778–1848). The three Zeiss brothers may have spent some time together in the city before Eduard, the eldest, became the rector of Buttstädt's local school in 1834. In January of 1835 Carl moved in with his brother Gustav in what was known as the *Körnerei*, a kind of student accommodation surrounded by inns that were owned by the master carpenter Körner.

A position at the University of Jena

From the summer semester of 1835 to the winter semester of 1837/38, Carl Zeiss was enrolled to study mathematics at the University of Jena. According to his university transcript, he attended lectures on algebra and analytical geometry, as well as experimental physics, trigonometry, stereometry, anthropology, mineralogy and optics. Alongside his focus on the geometrical disciplines relevant to the technical trades, Zeiss therefore also devoted his attention to areas of science that relied on scientific instruments. The field of optics that would be the focus of his subsequent career was added at a later date. The subject was taught by Friedrich Körner. As one would expect from someone studying while working, Zeiss opted for seminars and lectures that provided him with an introduction to different topics. Such a comparatively

14 Jenergasse as seen from Weigelstrasse, where Carl Zeiss lived for a time; taken in 1928.

low degree of specialization was also typical of scientific lectures as a whole; in the 19th century these still formed part of the Faculty of Philosophy along with history, poetry and classical philology. One example of this is the series of lectures on experimental physics taught by Jakob Friedrich Fries (1773–1843) that Zeiss attended in the winter semester of 1835/36. Fries was by his very nature a philosopher, but was also granted official permission to teach mathematics and physics.

Today, we cannot fully comprehend how a position like this was filled, but it did include a political dimension: on account of his closeness to radical democratic circles, Fries was forced to retire in 1819. The comparatively liberal atmosphere at the University of Jena enabled the devoted member of the fraternity and antisemite to find new ways to enter the world of academia. In 1824 Fries was granted permission to resume his teaching duties, though this was initially restricted to the 'unpolitical' realm of science. So it was that one of the pioneers of German *Gesinnungsnationalismus* (political nationalism) introduced Carl Zeiss to the basics of experimental physics.

Nevertheless, it is likely that Carl Zeiss spent the majority of his productive time in Jena at the workshop run by Friedrich Körner – in any case, he attended only one lecture each semester: Körner's residence and place of business, rather spacious premises, were located at Grietgasse 10, beyond the historic city walls, where Zeiss is purported to have lived for a time. The fact that Zeiss was involved in glass melting here means that there must have been an oven for this purpose on the premises in the 1830s – and

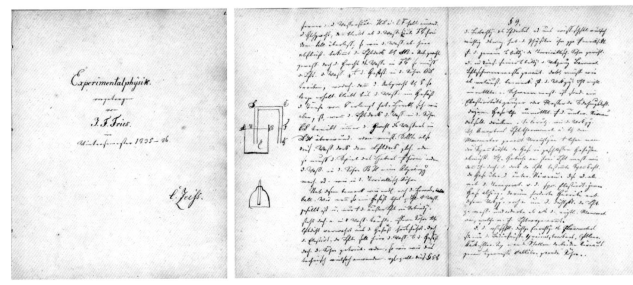

Carl Zeiss's lecture notebook pertaining to the lectures of J. F. Fries on experimental physics (winter semester 1835/36).

this was certainly a source of contention with the neighbors due to the immense amount of smoke it produced. But let us consider things one at a time: the parallels that can be drawn between Körner and Zeiss, who opened mechanical and optical workshops in Jena some 30 years apart, are interesting enough to warrant a closer inspection of Zeiss's predecessor and teacher. So who was Friedrich Körner, to what extent could he have been a role model for Zeiss and why did Körner end up a failed entrepreneur?

Körner and Zeiss

Friedrich Körner was born on 2 September 1778, the son of a baker from Weimar. In other words, just like Zeiss he came from a line of tradesmen employed by the Thuringian residence; bakers were some of the highest earners among ordinary citizens. Like Zeiss, Körner also went to Weimar grammar school, where he discovered his passion for mathematics and mechanics and chose to study instead of taking over his father's business. The sovereign Duke Carl August (1757–1828) of Saxe-Weimar-Eisenach was in favor of this decision and funded Körner's studies at the University of Jena. After graduating, Körner became a journeyman and probably worked at different mechanical and optical workshops. We do not know any more about this period of his life. In around 1810, Körner was employed as a mechanic at the Weimar Court with the assistance of Goethe, who held a position as a minister without portfolio. Two years later, he was tasked with building an 'air pump' (vacuum pump) for the chemist Johann

Johann Wolfgang Döbereiner. Print by Carl August Schwerdgeburth inspired by a Fritz Ries sketch.

Dr. Karl Dieterich von Münchow, inspired by a print from 1835.

Wolfgang Döbereiner (1780–1849). The instrument was first unveiled on 2 October 1812 in Jena before a large assembly of dignitaries. This prompted the Prussian government to order two similar pumps for the University of Bonn.

Körner's next great challenge was to help the director of the newly built observatory in Jena, Carl Dietrich von Münchow (1778–1836), equip his facility. The most prestigious plans pertained to the construction of a telescope with a diameter of 122 millimeters and a focal length of two meters. At the time, an instrument with dimensions like these constituted an immense challenge, especially since it was virtually impossible to find the glass suitable for producing its lenses. This is where one of Körner's traits comes to the fore: he had a tendency to fall headfirst into a subject even if he did not possess the relevant knowledge. Goethe, who was happy to support Körner,

remarked on his behavior rather flippantly: "If the man could listen as well as he can speak, he would be truly invaluable."[2] One way in which Carl Zeiss differed greatly from his employer was that he was very realistic and perhaps somewhat too cautious when it came to his abilities.

Körner reached his limits for the first time with the ambitious telescope project he worked on for the Jena observatory. Even though the instrument was completed in 1817 despite a series of technical difficulties, it was far too large to be used at the observatory. It later came to light that the French lenses that had been bought in the confusion stirred up by the Wars of Liberation were flawed to such a degree that the enormous and highly costly apparatus was suited as a museum exhibit at best.

Körner's house on 10 Grietgasse.

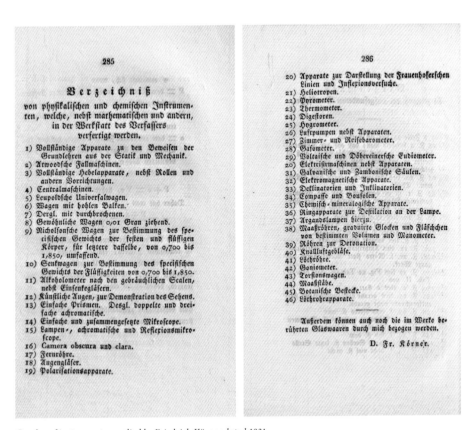

Catalog of instruments supplied by Friedrich Körner dated 1831.

Nevertheless, Körner continued to enjoy the benevolence of his sovereign. Carl August provided his court mechanic with an annual wage of 300 thalers, redeployed him to Jena and ensured that he completed a postgraduate degree in 1818 at the university and earned the qualifications required to work as a private lecturer without delay. Körner returned the favor by offering courses in optics and mechanics at regular intervals; he showcased the virtues of numerous instruments. He also founded a mechanical workshop in Jena with a portfolio focused primarily on the university's needs. While we can see that Zeiss turned his attention to building microscopes shortly after establishing his company some three decades later and saw all other activities such as trade in thermometers or spectacle repairs as additional

services, it does not seem that Körner had a core line of business as such. The *Verzeichnis mechanischer Arbeiten, welche in meiner Werkstatt angefertigt werden* published by Zeiss's teacher in 1824 enumerates 55 different product categories, among them a wide range of microscopes and telescopes as well as scales, meteorological instruments, electrostatic generators and artificial eyes.[3] The workshop was very modest in terms of its size. Körner was unable to afford a full-time workforce. Reference to the at times precarious financial situation in which he found himself is also made in the poem written to mark the construction of 10,000 Zeiss microscopes in September 1886. As the regional poet Leo Sachse put it rather ineloquently, Körner's workshop was "a place where even daughters partook in grinding."[4]

Cover page of an essay by Friedrich Körner, 1824.

Körner's fading popularity

Although Körner involved his entire family in the workshop, he was notorious for delays in his orders. In August 1824 a further delayed delivery that caused Körner to break his promises resulted in a quarrel with the Grand Duke Carl August. His long-time benefactor accused Körner of "being disobedient and guilty of highly improper conduct," suspended him from all his functions and stopped paying his wages. Körner was on the brink of ruin. His only option was to plead leniency for himself and his family. In October 1824 he wrote to the Grand Duke:

"The income I receive from orders placed from across the nation, and the even smaller local incomings, are not sufficient to survive on; the number of orders we receive from abroad is declining daily as a result of the pressure exerted by tolls […]. […] I now find myself obliged to seek out a new discipline that pays more handsomely than art, otherwise I will be forced to find lodgings elsewhere. It is not possible to venture down either of these avenues without jeopardizing my wife's meager fortune."[5]

But the court mechanic who had fallen from grace was soon able to heave a sigh of relief – after being reprimanded at length by Goethe, his privileges were restored. This episode does illustrate, however, that Körner was unable to break away from his financial dependency on his sovereign. The fact that Zeiss, unlike Körner, did subsequently not receive state funding, was undoubtedly a result of the changed economic climate. The establishment of the German Tariff Union (*Zollverein*), whose member states included Saxe-Weimar-Eisenach, meant that a large number of trade barriers were removed starting in 1834. Standardization was introduced for currencies, weights and measures. At the same time, demand for scientific instruments was constantly on the rise as a

Johann Wolfgang von Goethe, oil painting by Joseph Karl Stieler, 1828.

Utzschneider in Munich (at the site where Fraunhofer had formerly worked) cost less than half as much. Körner explicitly addressed "less-well-to-do connoisseurs,"[6] in other words price-conscious consumers, and used housings made of cardboard instead of the more commonly employed wood or brass tubes. This is another area in which Carl Zeiss differed greatly. He focused his business on quality and exports; we will delve into this aspect in greater detail later.

Pioneering genius

It would be wrong to view Körner's work as nothing but a mere precursor that eventually led to the subsequent achievements of his famous apprentice. In truth, Körner laid the foundation in more ways than one for the advances in optical construction that were to be achieved in Jena as the 19th century progressed. First and foremost, he created the theoretical basis for the field of optics. Körner collaborated with Jena-based mathematicians, joining forces first of all with Carl Dietrich von Münchow and later, when the latter moved to Bonn in 1826, with Friedrich Wilhelm Barfuss (1809–1854), who assisted the court mechanic in designing telescopes and microscopes. These collaborations did not result in any major breakthroughs – a fate that would subsequently be shared by Zeiss in the early years of his own career.

Körner invested even more energy in establishing his own glassworks. The court mechanic made his first attempts shortly after moving to Jena in 1817. This

result of the Industrial Revolution and the emergence of empirical natural research. All of these factors meant that, from 1846 onwards, Zeiss enjoyed far better conditions under which to establish an optical and precision-engineering business compared to those his teacher Körner had experienced in the 1820s.

Körner also attempted to gain market share through price dumping: in 1826 the Weimar-based court mechanic advertised telescopes produced at his workshop which, compared to those manufactured by may have been prompted by the difficulties associated with procuring glass from abroad to construct the apparatus used in observatories and the impressive feats achieved by Joseph von Fraunhofer (1787–1826) in the Bavarian town of Benediktbeuern. The Grand Duke Carl August was generous in his support of both the melting tests and the procurement of measuring instruments. Nevertheless, Körner's haplessness had almost tragic proportions. Upon first glance, his molten glass looked promising, but a series of very minor impurities and striae rendered it unsuitable for precision optics. The fact that Körner could not refrain from making grand promises would have done little to enhance his reputation in government circles. In May 1829 Goethe penned this entry in his diary: "Körner has acquired new glass samples and, as usual, is making a great fuss about them. Unsurprisingly, the results are inconclusive."[7]

From 1834 to 1838 when Zeiss was Körner's apprentice, he also took part in the melting experiments. It would seem that his teacher had not yet given up on his dream of taking glass production into his own hands. Körner made sure that his methods and formulas were kept under lock and key; Zeiss was enlisted solely for menial tasks. Perhaps it was this alchemist's furtiveness that stood between Körner and a breakthrough. Even so, it would be another 50 years before Abbe and Schott combined their expertise in optics and chemistry and gained a better understanding of glass.

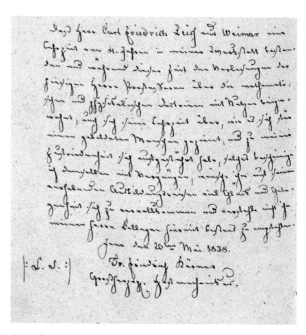

Copy of letter of reference written by Körner for Zeiss, 20 May 1838.

Journeyman years

On 20 May 1838, a Sunday, Körner released Zeiss into his journeyman years with a glowing reference. As mechanics was not a trade in its own right, there was no prescribed way in which to go about his new endeavor. Zeiss was left to fend for himself. The only autobiographical account from this period in the young mechanic's life dates back to 1846. In his search for a place to set up his business, Zeiss wrote the following:

"In the period from January 1838 to May 1845 I sought to further my knowledge by working at the

most renowned physical, optical, mathematical and machine workshops in Stuttgart, Darmstadt, Vienna and Berlin. I took every opportunity for my advancement by using the tools and acquiring the skills beneficial to and necessary for a mechanic." [8]

Few historical records exist of his journeyman years. This is not surprising when we consider that the arrival of a lone, and at the time unknown, wandering mechanic was an unremarkable event in a big city. The large gaps in his employment and university references would also suggest that as a journeyman, Zeiss did not solely focus on advancing his knowledge and skills. In light of the fact that Zeiss was a young man in his twenties who had escaped the provincialism of Weimar and Jena for the first time this is all very understandable.

There are no records about the time Zeiss spent in Stuttgart; the same goes for Darmstadt. Different researchers have speculated that Zeiss spent some time in Darmstadt working at the factory of Johann Hector Roessler (1779–1863) which primarily produced minting machines. This is plausible: Roessler was a journeyman in Jena, assumed the post of a university mechanic in Darmstadt and before being appointed Grand Ducal Currency Officer for Hesse in 1832 also ran a mechanical workshop that was not dissimilar to that of Körner in Jena. Nevertheless, there is no documentary evidence to support the claim that Zeiss was employed by Roessler in Darmstadt.

A mechanic in Vienna and Berlin

The last two stops during his journeyman years offer a somewhat clearer picture. There are two accounts of his time in Vienna: the first from the Imperial and Royal Polytechnic Institute, the nucleus of the Vienna Technical University. In 1842/43, Zeiss attended "the Sunday lectures on popular mechanics, diligently" [9] and passed the examination with a very high grade. The second reference comes from the Viennese branch of the Strasbourg-based railway and carriage scale factory Rollé & Schwilgué, which operated from 1844 onwards as Heinrich Daniel Schmid (following multiple name changes, it is now part of Siemens AG Austria). Zeiss received documented confirmation of his time spent working at this particular Vienna-based heavy machinery producer between April and August 1843. Yet it would seem that the young mechanic did not leave the city immediately there-

Poststrasse in Berlin, 1888. Photograph by F. A. Schwartz.

Hot-air balloon ride over Vienna, 1847. Watercolor by Jakob Alt.

after. A good decade later Zeiss wrote to his friend K. O. Beck to inform him that he did not leave Austria until June of 1844 and then made his way to Berlin via his hometown of Weimar.[10] This fits with the reference Zeiss received from the Berlin-based mechanic C. Lüttig. Zeiss then worked as an assistant in the workshop on Poststrasse 11 in Berlin's Nikolaiviertel district from 30 September 1844 to 6 September

1845. As Lüttig also took part in the General German Trade Exhibition in the Prussian capital in 1844, we know that he specialized in technical drawing instruments for physical and mathematical research.[11]

Given the way his career would turn out, it is surprising how little Zeiss concerned himself with optics during his journeyman years. Not one of the references he received was issued by an employer specialized in optical instruments, although the many ports of call along his route included Simon Plössl in Vienna and Friedrich Wilhelm Schiek in Berlin who ran workshops that enjoyed excellent reputations. It is clear that Zeiss was interested primarily in mechanics; he would not discover the microscope until later. This would not happen before he returned to Jena and encountered the ambitious team of nature researchers led by the botanist Matthias Jakob Schleiden (1804–1881).

While Zeiss had been traveling far and wide, Schleiden had already managed to persuade the court mechanic, Körner, to devote more time to the construction of microscopes. By now, however, Körner was already over 60 years of age and could not keep pace with Schleiden's passion for research. Perhaps the time had come to look for a younger mechanic who was prepared to once again push the boundaries of optical instrument design.

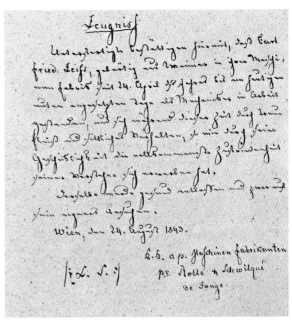

Copy of letter of reference from Rollé & Schwilgué in Vienna, 24 August 1843.

1 Cited by: Hans-Werner Hahn: "Zwischen ständischer und bürgerlicher Gesellschaft" in: *Ereignis Weimar-Jena, Kultur um 1800*, ed. by Olaf Breidbach, Klaus Manger, Georg Schmidt. Paderborn, 2015, p. 37.

2 Cited by: Herbert Koch: "Neues über den Hofmechanikus Dr. Friedrich Körner" in: *Zeiss-Werkzeitung Nr. 3* (1936), p. 63.

3 Moritz von Rohr: "Zu Friedrich Körners Gedächtnis" in: *Deutsche Optische Wochenschrift Nr. 5* (1927), p. 55.

4 Cf. Horst A. Willam: *Carl Zeiss. 1816–1888*. Munich, 1967, p. 126.

5 Cited by: Koch, *Neues*, p. 63.

6 Cf. Moritz von Rohr: "Weiteres zu Friedrich Körners Gedächtnis" in: Deutsche Optische Wochenschrift Nr. 51 (1927), p. 701.

7 Cited by: Koch, Neues, p. 63.

8 ZEISS Archives, BACZ 13893.

9 Reference from the Imperial and Royal Polytechnic Institute dated 16 July 1843. Source: ZEISS Archives, BACZ 13893.

10 Letter from Carl Zeiss to K. O. Beck dated 4 February 1855. Source: ZEISS Archives, CZO-S 3.

11 Cf. *Ausführlicher Bericht über die große allgemeine deutsche Gewerbe-Ausstellung in Berlin im Jahre 1844*, ed. by Amand. Ferd. Neukrantz. Berlin, 1845, p. 285.

Obstacles to Setting Up a Business:
Founding of the Mechanical Workshop in Jena

After eleven years of being an apprentice, Carl Zeiss returned to his home town of Weimar in early September 1845. What happened next amounts to one of the many gaps in the biography of the man who would later become the company founder. In 1921 a report in the Zeiss plant newspaper discussed a statement made by a certain Karl Hindersinn, according to whom Zeiss wanted to establish a business as a lensmaker and mechanic in Weimar after having spent time in Berlin. His entreaty, however, is said to have been quashed by the authorities. According to the text, Zeiss also "inveighed against the Weimar authorities, gathered his worldly possessions and made his way to Jena, his vise slung over his shoulder."

Recent investigations conducted by the ZEISS Archives show that this is highly implausible. Nevertheless, this episode is so very memorable and encapsulates the contradiction between the old and musty Weimar and the all-liberal Jena so aptly that it became something of a legend that was part and parcel of all Zeiss biographies as of the end of the 1960s at the latest.

There is no actual evidence that Carl Zeiss registered his business in Weimar. No reasons can be found to explain the immediate rejection of any such entreaty by the Weimar authorities, especially since as the godson of the Grand Duke and the son of a respected craftsman of the court, Carl Zeiss was not exactly a

Holzmarkt in Jena circa 1900.

nobody. Moreover, the authorities would have had very little time to process the entreaty as Zeiss came to Jena on St. Michael's Day (29 September) in 1845. For 29-year-old Zeiss, who had just completed his journeyman years, setting up a business in the small university town of Jena may well have been a more attractive prospect than going back to work at his father's business, which would have meant moving back into the family home with all the limitations that presented. Another factor in Jena's favor was that it offered a propitious environment for setting up a precision-engineering business given the value accorded to science there.

Student status paves the way to a new life

Carl Zeiss, who was familiar with the city as a result of his apprenticeship under Friedrich Körner, rented a room on 27 October 1845 in the home of the Schoemanns at Holzmarkt square 16, less than 100 meters away from Neugasse 7, where he opened his first workshop in 1846. But before he could even begin to consider establishing a business, Zeiss needed a residence permit – and proof of abode was insufficient to obtain one. The simplest way to meet these requirements was to enroll at the university. And so it was that Carl Zeiss became a student for the second time in his life. On 30 October 1845 he was granted permission to attend lectures on mathematical analysis.[1] On this basis, Jena's police department issued Zeiss with a one-year residence permit on 3 November. His sister Pauline helped the young mechanic put

Portrait of Matthias Jakob Schleiden by J. H. Schramm, circa 1845.

down roots in the city he could once again call home. It would seem that she kept house for Carl until he married in 1849. His younger sister Hulda is also said to have helped her brother at the time.

Initially, Zeiss postponed his plans to open his own workshop. There is no documentary evidence to suggest why he did this. We can assume that Zeiss needed compelling arguments as to why a further workshop should be set up in Jena given that his two potential competitors, the court mechanic Friedrich Körner and the university mechanic Johann Friedrich Braunau were already firmly established in the city. A sufficient reason could have been found had Körner or Braunau decided to dissolve their business. This would not have come as a complete surprise in the case of Körner, who was already 67 in 1845. Zeiss was well aware of this and years later admitted in a letter to a friend from his time in Vienna, K. O. Beck, who was working in Moscow, that Körner was "old and haggard" in the period in question and was no longer able to meet the needs of scientists in Jena.[2] But Körner had not yet retired and if Zeiss did not want to wait until the opportunity to step in presented itself, he had no other choice but to demonstrate beyond a shadow of a doubt that there was a veritable need for another workshop. In light of this situation, it was only logical that while attending lectures he also began working at the Institute of Physiology established in 1843. The Institute was run jointly by the botanist Matthias Jakob Schleiden, the geologist Ernst Erhard Schmid and the physician Heinrich Haeser.[3] Under the aegis of a private research facility, the Institute focused on empirical work, which necessitated a large amount of scientific apparatus.

Zeiss jumps through hoops

Zeiss was already in close contact with key potential customers but he had yet to receive a permit to run a precisionengineering workshop. On 10 May 1846 he submitted a request to obtain one from the Grand Ducal provincial headquarters. In it, Zeiss gave an account of his career and made reference to his extensive "training in the most renowned physical, optical, mathematical and machine workshops in Stuttgart, Darmstadt, Vienna and Berlin" before providing details of how he planned to establish his business:

"As when establishing a production facility for advanced mechanics due consideration must be given from the outset to creating all conditions required to ensure a comprehensive and expanding business, and since a direct link to the men of science lays the best foundation to accomplish this objective, it would seem to me that in our Grand Duchy the university city of Jena is the most opportune place of establish-

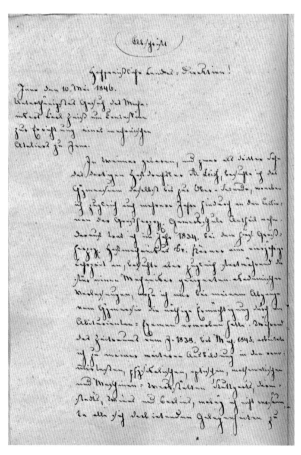

First page of the copy of the request put forward by Carl Zeiss to the provincial headquarters on 10 May 1846.

ment, and all the more so because teachers and students of the university require a constantly increasing amount, but this need certainly cannot be satisfactorily met locally at the present time […]. – […] I shall therefore allow myself to make the most humble request before I take any further course of action:

I hereby entreat the highly esteemed provincial headquarters to kindly accord me the necessary license to establish a workshop and, respectively, the commercial business for all articles relevant to mechanics, as well as for all other instruments employed for physical and chemical purposes in the Grand Duchy of Weimar and especially for the city of Jena."[4]

Three primary aspects stand out: first of all, when setting up his workshop Carl Zeiss certainly had no intention of focusing on optical production. He wanted to offer scientific apparatus that bore a direct correlation to his training as a mechanic. Second, the intention to found a further workshop in Jena was justified in that the existing suppliers were unable to keep pace with the growing demands of the scientific community. As such, the company Carl Zeiss is closely associated with the rise of sciences based on empirical research. Third, Zeiss had already been thinking about a "comprehensive and expanding business" even before he moved into his first workshop, in other words he was thinking about exporting his products beyond the limits of the city and the country.

To give weight to his request, Zeiss included a series of references and letters of recommendation from the professors Schleiden, Schmid and Haeser of the

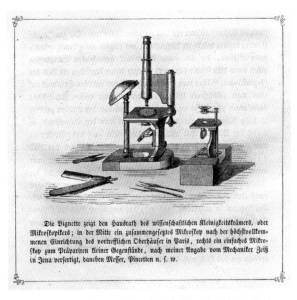

Die Vignette zeigt den Hausrath des wissenschaftlichen Kleinigkeitskrämers, oder Mikroskopikers; in der Mitte ein zusammengesetztes Mikroskop nach der höchstvollkommenen Einrichtung des vortrefflichen Oberhäuser in Paris, rechts ein einfaches Mikroskop zum Präpariren kleiner Gegenstände, nach meiner Angabe vom Mechaniker Zeiß in Jena verfertigt, daneben Messer, Pincetten u. s. w.

Excerpt from "Die Pflanze und ihr Leben" by Matthias Jakob Schleiden, Leipzig, 1855, p. 29.

Physiological Institute. Schleiden in particular made every effort to support Zeiss's plans to set up a business and, like Zeiss, based their necessity on the growing global demand for usable instruments:

"If I consider the situation from a commercial standpoint, I do not know how I am to justify the need for a third mechanic in Jena given that I would not know how to justify the existence of the first and second […] ones. People in a big city like Paris, Vienna or Berlin may well place orders for items to meet everyday requirements, or indeed to satisfy the need for luxury goods that have now become commonplace, in quantities large enough that one or more mechanics could make a living.

[…]

But this perspective is […] wholly incorrect, especially in our times. […] The mechanic and the lensmaker are […] regarded in the same way as an artist. But who would even dream of preventing a landscape painter, a historical artist, or the like, from conducting their work on the grounds that a specific place has no need for it. For an artist, the world is the audience and this is where the artist can be likened to the mechanic.

[…]

Of all the many microscopes, for instance those found here in Jena, apart from the small microscopes produced by Dr. Körner […] there are none that have not emerged from workshops in Vienna, Berlin or Paris. Much the same can be said of countless other commodities […]. All of the well-known mechanical and optical workshops in Germany together are unable to respond to the daily increase in demand, and the general source of discontent among all nature researchers is that they are compelled to wait for years before they receive what they have ordered."[5]

In addition to the aforementioned university professors, Jena's municipal council was also in favor of Zeiss's plans to found a business. In a letter dated 10 June 1846, the provincial headquarters announced that Zeiss would be granted community rights as soon as he received a business permit.[6] In spite of his excellent references, the provincial headquarters went through all the proper channels and initially referred the case to the Grand Ducal Building

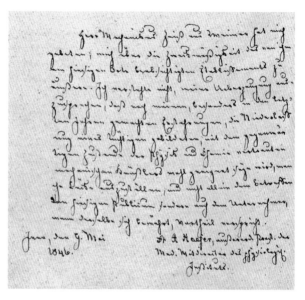

Copy of the letter of recommendation written by Haeser (9 May 1846).

Authority, which in turn summoned Zeiss to Weimar to assess his professional suitability in more detail. In August 1846, Zeiss conducted a total of 13 assignments over a four-day period. The mathematical and technical knowledge required for the examination questions was well above the level one would associate with the term "mechanic" today. In his responses, Carl Zeiss also demonstrated that his mathematical and linguistic skills bore more resemblance to the qualifications of a modern-day engineer.

A hint of rebellion: Zeiss sabotages his exams

Zeiss made no secret of the fact that he considered the questions as nothing more than an impertinence and a waste of time. Instead of responding to them as efficiently as possible, he railed against the examining authority. The questions, he said, "lacked the

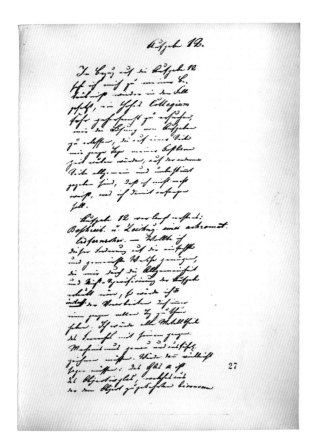

Excerpt from the examination paper taken by Carl Zeiss, August 1846.

mathematical accuracy and precision […], which are essential in matters of applied mathematics." He refused to provide solutions to other questions on the basis of the sheer time and effort this involved and instead referenced his earlier work. Zeiss furthermore gave vent to his displeasure with passing remarks that sometimes seemed to take longer than the explanations pertaining to the task at hand. From a biographer's perspective, these comments that could have cost Zeiss his permit were not only entertaining but also provided a very real impression of the 'start-up' entrepreneur's quick temper. For example, the examiners' request for a "description and drawing of a functional achromatic terrestrial telescope" was met with the following response from Zeiss:

"In relation to question 12, I am regrettably compelled to respectfully request the esteemed committee to refrain from asking me to provide solutions to questions that would occupy days of my precious time and are also formulated in a way that is far too general and imprecise and leave me at a loss as to what specifically is being asked of me."[7]

In his response to the following and final question, Zeiss lashed out at the examining authority with these concluding remarks:

"I was of the opinion that it would please an esteemed council to conduct the examination orally. But this is the fourth day I have spent here to provide written responses to questions and tasks put to me by the esteemed committee under their supervision.

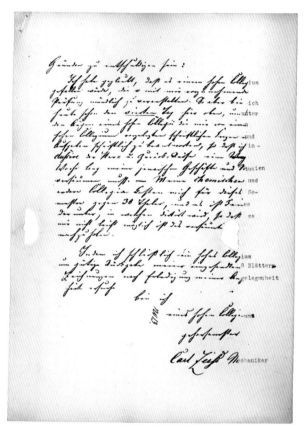

Excerpts from the examination paper taken by Carl Zeiss, August 1846. Part of the last page was later reconstructed in typewritten from.

As a result, as well as having to travel to and from Jena, I am unable to run my business and continue my studies for an entire week. My chemistry and other lectures are costing me some 30 thalers this semester, and as none of the content is dictated, it is not easy for me to make up for what I have missed in my absence. By finally entreating the esteemed colleagues to afford me a favorable return for the 8 sheets and drawings I submitted after bringing my affairs in order, I am, sincerely,

Carl Zeiss, mechanic to an esteemed committee."[8]

The building inspector in Weimar, Carl Georg Kirchner, was entrusted with providing an expert assessment. In the 30-year-old Zeiss's responses, he saw all the knowledge and skills necessary for setting up a workshop, but found fault with the candidate's less-than-constructive attitude and rightly said: "Mr. Zeiss sees […] difficulties where in fact there are none."[9] Weimar's Building Authority now faced a dilemma: while there was no reason not to grant a permit based on the facts, the offense caused by Zeiss's disrespectful attitude could not be ignored. The authorities did what they felt they had to: absolutely nothing.

Raring to go

Carl Zeiss became ever more impatient as the months went by and had still received no word from Weimar. On 21 October he wrote to the Jena municipal council and asked that they inquire as to the status at the Grand Ducal provincial headquarters. This came to pass on 24 October and the provincial headquarters

First workshop on 7 Neugasse, sketch from 1896.

Concession certificate of the Grand Duke of Saxe-Weimar-Eisenach for the mechanic Carl Zeiss of Jena "to produce and sell mechanical and optical instruments and to establish an atelier for mechanics in Jena," 19 November 1846.

then approached the Building Authority on 27 October. Only then did the buildings officer in charge, Heinrich Hess, readdress the issue and responded on 7 November 1846, still somewhat piqued:

"With this letter we kindly inform you that the mechanic Carl Zeiss of Jena has, with a satisfactory

A photograph from the garden of the house on 7 Neugasse taken between 1870 and 1890 paints a different picture from that of the reconstruction sketch from 1896, which is used most often. Thank you very much to Falk Burkhardt for this information.

grade, passed the examination he took as a result of your honored letter dated 13/18 June of this year to assess his knowledge and acumen on the subject of mechanics. It would nevertheless have been preferable for [Mr.] Zeiss to have refrained from being misled by his inherent self-confidence in his own skills to transcend the limits of the modesty befitting to providing written answers to the questions."[10]

From that point on it was plain sailing: on 19 November 1846 the provincial headquarters in Weimar granted Zeiss a permit and notified the Jena municipal council; on 26 November Zeiss received a letter from the latter, and on 8 December Zeiss officially became a citizen of Jena. It would seem, however, that he was already unofficially aware of the news.

1 Cf. Koch, *Unbekanntes*, p. 1.

2 Letter from Carl Zeiss to K. O. Beck dated 4 February 1855, ZEISS Archives, CZO-S 3.

3 Cf. also Joachim Wittig: "Carl Zeiss und die Universität Jena" in: *Carl Zeiss und Ernst Abbe*, ed. Rüdiger Stolz, Joachim Wittig. Jena, 1993, p. 23.

4 ZEISS Archives, BAZC 13893, sheet 71.

5 ZEISS Archives, BAZC 13893, sheets 76–77.

6 ZEISS Archives, BAZC 13893, sheet 78.

7 ZEISS Archives, BACZ 11347.

8 ZEISS Archives, BACZ 11347.

9 ZEISS Archives, BACZ 11347.

10 ZEISS Archives, BACZ 11347.

Selling Like Hot Cakes:
Zeiss Builds His First Microscopes

As early as 17 November 1846, a Tuesday two days before the permit was officially granted, Carl Zeiss is reported to have moved into his first workshop on Neugasse 7. This is now regarded as the date on which Carl Zeiss AG was officially founded but it is not supported by any documentary evidence. The company Carl Zeiss was not entered into the Jena commercial register until 16 April 1863.[1] The original date of establishment was later reconstructed on the basis of the workshop logbook dating back to 1871; in it, the mechanic Pape wrote the following note under 17 November: "Stopped at 9 a.m. Company's

25th anniversary."[2] There is also a photograph from 1871 depicting the then 19-strong workforce. It bears the same date.

What we can say for certain is that Carl Zeiss was on thin ice when he made his first forays into setting up his own business: his brother Eduard, principal of Jena's Citizens' Schools, gave him 100 thalers to help him set up his workshop.[3]

House on Wagnergasse in Jena, second workshop of Carl Zeiss; photograph from 1906.

This amount, which was later reimbursed by his father, would equate to some 3,000 euros by today's standards.[4] In the first six months that followed the founding, there were no notable business activities to report. Carl Zeiss wrote a retrospective letter to his friend Beck in which he said that he was still working on equipping his business in February 1847.[5] It was not until 5 May 1847 that an advertisement appeared in the newspaper *Privilegirte jenaische Wochenblätter* announcing Carl Zeiss as a supplier of optical products:

"I now stock spectacles, and botanical and other loupes. In a few days I will also be supplying a selection of inexpensive thermometers."[6]

Optical stores were not yet commonplace. Even spectacles were often still being sold by traveling

Eduard Zeiss, Carl Zeiss's brother.

"While I was there, I saw that he had finished constructing a microscope; I found it very good indeed. He does not wish to build any more of them, however, until he receives the machine from Berlin that is due to arrive this month."[7]

The machine mentioned in the letter was a lathe produced by the engineer August Hamann in Berlin and without which efficient microscope production would have been unthinkable. Assuming that Zeiss received the machine on time, he would have been able to begin producing microscopes in the summer of 1847. He sold his first instrument in September 1847 to Hermann Schacht, a botanist from Hamburg who had studied under Schleiden and worked for a time in Jena before accepting a position as a professor in Bonn.

Schleiden sets the pace

Zeiss's decision to turn his attention to microscope production early on, instead of running a general store like many of his fellow mechanics, is probably due to his close contact to the head of the Physiological Institute, Matthias Jakob Schleiden. A follower of Charles Darwin's work and co-founder of the cell theory, Schleiden was hailed as a pioneer of microscopy. At the beginning of the 1840s, Schleiden encouraged the Jena-based mechanic Körner to produce simple microscopes. From then on, Schleiden checked each microscope Körner produced before it was delivered. In February 1847, however, Körner passed away and now the man who was probably the most

Optical lathe circa 1900.

salesmen. The workshop on Neugasse did not seem to be much more than a temporary location and on 19 June 1847 Zeiss published an announcement in the paper that he would henceforth be living and working at Wagnergasse 32. A letter from his father, who assisted him with the move, tells of a prototype for a microscope:

well-known scientist in Jena was in desperate need of a new microscope-maker to cover the needs of his Institute and the teaching of his students.

Schleiden provided a detailed account of the status of microscope construction in his 1842 book *Grundzüge der wissenschaftlichen Botanik*. After having scrutinized the best models on the market in both Germany and abroad, he came to quite a sobering conclusion:

"Concerning the setup of the microscope, I must agree with Hugo von Mohl [the botanist from Tübingen] that the way in which instruments have been produced until now does not wholly match any of the requirements of the user [...]. The main requirements are as follows: coarse and fine movement, both with respect only to the body of the microscope; the stage, fixed in position with an aperture of approximately 1/2 inch in diameter, under which is a rotatable perforated disc; a plano-convex illumination lens with a focal range of approximately 1.5 inches and a flat mirror that can be laterally adjusted [...].

Furthermore, it would also be meaningful to completely omit the high-powered eyepieces [...] as they are wholly unusable and unnecessarily raise the price

Matthias Jakob Schleiden from the book "Studien: Populäre Vorträge," 1855.

Watercolor by Schleiden of his house on 10 Neugasse in Jena (Photo: G. Uschmann).

of the instrument. – Every microscope observer will, in the end, find a large number of small apparatus, bad forceps, small blunt scalpels, cover slips for infusoria and other such useless items, things that have been there for so many years, just where the producer left them because they are entirely useless; these are the sorts of items that people should finally start disposing of."[8]

It would seem that Carl Zeiss put this critical account to good use when building his first microscopes in 1847. In fact, Zeiss attempted to largely incorporate the enhancements outlined by Schleiden. This is demonstrated by the first advertisement we know of for Zeiss microscopes, which appeared in two editions of the *Augsburger Allgemeine Zeitung* in September and October 1847:

"The undersigned herewith announces to persons conducting research into nature that he will henceforth stock small microscopes, known as doublets. All of the latest requirements from physiologists have been incorporated into such constructions. And the stage is in fact fixed in position; the movement of the microscope is possible by a shifting action, and for finer settings using a screw with a spring, resulting in adjustment that is fast and convenient for lower magnifications and also very fine and totally backlash-free for higher magnifications; an additional converging lens for higher magnifications is included with the illumination mirror. Aside from the necessary specimen and cover slips, three lens combinations are included; they permit linear magnification of 15, 30, and 125x.

Dissection microscope with dovetail mount – produced in this form until 1848 – from the collection of Timo Mappes; photograph taken by Manfred Stich.

The entire instrument has been placed in a box made of polished walnut wood and arranged in a way such that it can be set up more quickly and more safely, not in the usual manner by screwing action, but rather by using a suitable mechanism. The apparatus in its entirety costs 11 thalers. Clients are requested to send their orders by mail. Payment shall be made in advance by mail provided no other conditions have been agreed upon beforehand."[9]

Under the advertisement, Zeiss also included a detailed recommendation from Schleiden, who praised the technology and, most of all, the availability of the microscopes:

"I can highly recommend the microscopes offered by Mr. Zeiss in every way [...] and attest that they fulfill their purpose. [...] At the assembly of the northern German Association of Pharmacists that recently took place in Jena, all stocks of these [...] instruments were immediately depleted. I would add that Mr. Zeiss now stocks so many instruments that every order he receives is dispatched immediately; this is not a service one finds everywhere."[10]

At the time, Zeiss had sold exactly six microscopes. As previously mentioned, the first went to Hermann Schacht, two others to pharmacists in northern Germany, another to a professor of medical science, one to a student from Jena and the sixth was purchased by Schleiden. By the end of his first year in business, Zeiss was to sell a total of 29 instruments, most of them to buyers outside of Jena.

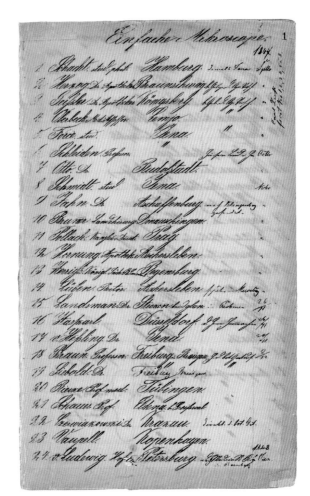

The first page from the book detailing microscope deliveries.

1   Cf. Axel Stelzner: "Carl Zeiss in der Jenaer Tagespresse (1847–1888)" in: *Carl Zeiss und Ernst Abbe*, ed. Rüdiger Stolz, Joachim Wittig. Jena, 1993, p. 108.

2   Cf. e.g. Friedrich Schomerus: *Geschichte des Jenaer Zeisswerkes 1846–1946.* Stuttgart, 1952, p. 10, note 12.

3   Cf. *Erich Zeiss, Hof- und Universitätsmechanikus*, p. 27.

4   Cf. *Kaufkraftäquivalente historischer Beträge in deutschen Währungen*, accessible at: www.bundesbank.de (Status: 15 January 2015).

5   Letter from Carl Zeiss to K. O. Beck dated 4 February 1855, ZEISS Archives CZO-S 3.

6   Stelzner, *Tagespresse*, p. 100.

7   Erich Zeiss, *Hof- und Universitätsmechanikus*, pp. 27–28.

8   Matthias Jakob Schleiden: "Grundzüge der wissenschaftlichen Botanik (1842)" in: *Wissenschaftsphilosophische Schriften*, ed. Ulrich Charpa. Cologne, 1989, p. 134.

9   Horst Alexander Willam: *Carl Zeiss. 1816–1888.* (Tradition, supplement 6) Munich 1967, p. 35.

10 Ibid.

Chapter 3

He Who Dares, Wins:
Zeiss Establishes His
Company

≪ *Page 49: Carl Zeiss and microscope with horseshoe stand circa 1861.*

Diploma from the 2ⁿᵈ General Thuringian Trade Exhibition in Weimar, 23 July 1861, for the presentation of the first honorary prize.

"The Best Simple Microscopes": Setting Up the Business (1847–1859)

Even before he sold his first microscope in September 1847, Carl Zeiss had already started looking for an assistant. With such limited capital at his disposal, however, he could hardly expect much of the potential applicants. He therefore aimed his job advertisement – published in the local newspaper *Jenaische Zeitung* on 7 August 1847 – squarely at unskilled workers:

"I am currently seeking a poor, honest young man aged between 14 and 16 to help me with some simple tasks in my business on a long-term basis for a weekly wage."[1]

Fortunately for Zeiss, this single advertisement proved to be sufficient to attract someone who would go on to play a key role in the company's success: August Löber (1830–1912). The son of a

craftsman, Löber was already 17, and therefore theoretically too old to apply, but the death of his father in January 1847 had left him in a position of genuine hardship. There is no record of whether this influenced Zeiss's decision to employ him, or indeed whether anybody else even applied for the job in the small town of Jena. But what we do know is that over the next 44 years Löber was promoted from apprentice to foreman, eventually ending up as the head of production and the company's most valued instructor and trainer. Moritz von Rohr tells us that Löber received a share of the profits later on in his career and became a wealthy man.[2] This illustrates the extraordinary regard in which the foreman was held in the company, even though much of the workforce considered his managerial style to be short-tempered and authoritarian.[3] Ernst Abbe lauded the

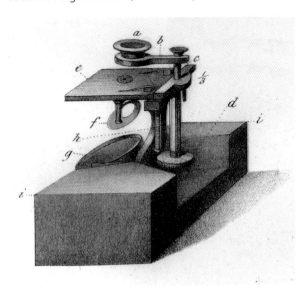

Simple dissection microscope produced by Zeiss; drawn by Hermann Schacht, from his book "Das Mikroskop," 1855.

August Löber, the first apprentice, who later became foreman, circa 1870.

August Löber with mechanics and apprentices in 1864 (from left to right): Carl Müller, Friedrich Pfaffe, Joseph Rudolph, Wilhelm Böber, Heinrich Pape, Fritz Müller and August Löber.

first 'Zeissian' in a speech he gave to commemorate the 50th anniversary of the Jena Optical Works:

"We are all delighted to still have him among us, our dear old August Löber, the founder of our school of ingenious technology, the senior member of our cooperative and the master teacher – directly or indirectly – of all our hard-working lensmakers. The success Zeiss has enjoyed was considerably buoyed by the fact that the first employee he hired to help him pursue his goals was able to offer such a willing appreciation of the peculiar nature of the tasks involved, such a highly developed sense of precision and exactness and such utter dedication of his whole being."

Weapons for the militia in 1848

Despite this stroke of good fortune with Zeiss's first employee, the commercial success of ZEISS was by no means assured in those early years. The March Revolution in 1848 had a dramatic impact on the business, as August Löber recalled:

"Since there were times when lensmaking work was scarce, I was occasionally compelled to help out with mechanical work. In 1848, Mr. Zeiss was a member of the local militia. There was little to do in the workshop, so we began making percussion locks out of old flintlocks; there were hammers to file

and hardening to be done. Those weren't the only difficult years – in the 1850s the business was beset by so many economic woes and price increases that we had to dismiss the second assistant and only Mr. Zeiss and my humble self remained as the entire workforce. […] From all this it would be fair to deduce that there were often few morsels left in our 'Egyptian flesh pots.'[4] I recall, for example, that Mr. Zeiss would breakfast simply on a roll and a small glass of Korn for 3 pfennigs, something I saw with my own eyes, occasionally receiving a sip of the schnapps myself. […]. It was not unusual for me to be summoned from my Sunday occupation (gardening) to work on a shoddy pair of spectacles for 1.80 marks. It therefore comes as little surprise that I never suffered from obesity."[5]

It is worth delving deeper into Löber's somewhat idiosyncratic musings. First of all, we discover that during this period of revolution Carl Zeiss was a member of the Jena militia, which temporarily took on the role of a military and police force at a time when the same was happening in most other German cities. The act of citizens taking up arms was on the one hand a liberal protest and a sign of resistance against the feudal authorities. But it was also an attempt to protect public order and, in particular, private property against criminal radical political attacks such as those undertaken by groups of students. His involvement in the militia therefore demonstrates that Zeiss was on the side of the moderate elements in the revolution. Löber's assertions that they "made percussion locks out of flintlocks" simply means that the Zeiss workshop was engaged in modernizing small firearms by

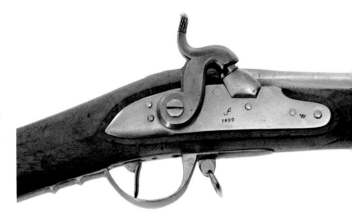

Percussion lock in which an impact-sensitive ignition charge replaces the flint in a flintlock musket. Photograph by Christian Stoye.

converting them from a flintlock into a caplock mechanism. The "hammers" Löber refers to also form part of a handgun's trigger mechanism.

A glance at the ledger Carl Zeiss kept from March 1848 onwards – something he referred to as the 'Manual' – confirms Löber's recollections. If we look at the month of October, for example, when Jena was occupied by Saxon troops as part of the German Confederation's intervention to put down radical democratic forces in Thuringia, several firearm-related jobs appear under the names of local citizens, including entries such as "rifle barrel bronzed," "sight blue annealed" and "applied percussion cap". Even Zeiss's long-standing patron Matthias Jakob Schleiden, whose membership of the Academic Association for Reform clearly identifies him as part of the progressive elements[6], ordered a powder horn instead of another microscope. Optics manufacturing did not come to a complete standstill, however. The workshop also deliv-

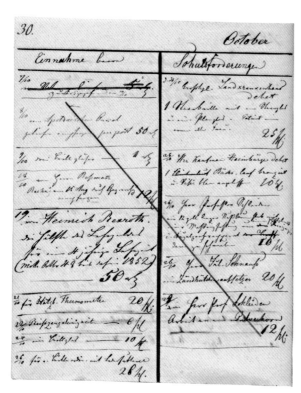

The 'Manual' belonging to Carl Zeiss. Page from October 1848.

ered a pair of "cataract spectacles" which had high minus power lenses to compensate for the specific vision problems resulting from cataract surgery. The precarious nature of Zeiss's finances at the time is illustrated by an entry which appears on 8 October: "borrowed money from uncle in Apolda, received 50 [thalers] by mail." Another indication of difficult times is a note in the Wages section which states that Mr. Diedrich, who was employed by Zeiss as a second assistant in addition to Löber, received his "last wage" on 7 October. From this point onwards until November 1852, Löber was the sole wage earner listed in the 'Manual.'

Identity crisis

Zeiss was clearly struggling to keep his business afloat, and his troubles were compounded by the stagnant development of his core microscope business:

Year	No. of microscopes sold[7]
1847	23
1848	38
1849	28
1850	22
1851	38
1852	38

Löber's words cited above imply that money was too short even to buy food. This posed an unfortunate predicament for Carl Zeiss, who was in the process of starting a family. Records suggest that his elder brother's wife introduced the young mechanic to the family of a clergyman named Schatter who lived in the town of Neunhofen in Thuringia. On 29 May 1849 Zeiss married the clergyman's daughter, Bertha Schatter. Years later Zeiss confided in his friend K. O. Beck that he had made a good choice in his union with Bertha, even though the bride had virtually no wealth of her own.[8] The couple's wedded bliss was short-lived, however: Bertha died the day after giving birth to her first son Roderich on 23 February 1850. She was just 22 years old. We shall never know whether this death in childbirth could have been avoided if the doctor had been sober when he was called from a masked ball to her bedside at the onset of labor.[9]

Fortunately Zeiss was able to call on family assistance once again. Bereft of his mother, Roderich was initially taken in by Zeiss's parents-in-law in Neunhofen. Therese Schatter looked after her newborn grandson until she died in February 1851, almost exactly one year after her daughter. She left Roderich a sum of money that was set aside to pay for his education. Following the death of his mother-in-law, Zeiss entrusted Roderich's care to his second-eldest sister, Hulda, who also seems to have spent much of her time in Neunhofen. This suggests that Carl Zeiss was entirely spared the duties of a single father, allowing him to focus all his attention on his business in Jena. Nevertheless, the young entrepreneur was clearly affected by his professional and personal tribulations.

He considered escaping his plight by starting a new career. On 18 August Zeiss wrote to the University of Greifswald and offered to relocate to the Baltic Sea if the Faculty of Philosophy, which was in charge of sciences, would be willing to appoint him as the supervisor of the faculty's physical instruments and pay him a fixed annual salary for his services as a mechanic.[10] He appeared to have chosen a favorable moment since Greifswald had been left without any

qualified person to do this job when their previous mechanic Friedrich Adolph Nobert moved to Barth. Zeiss eventually received a reply in October 1850, but it was not what he had hoped for. The economist Eduard Baumstark, who was the Dean of the Faculty of Philosophy at the time, informed Zeiss that no decision could be made unless Zeiss was

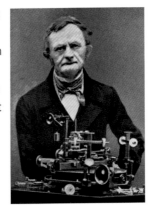

Friedrich Adolph Nobert (1806–1881), mechanic and optometrist.

first willing to fabricate a number of complex trial instruments, all of which would naturally be unpaid.[11] Even if he were to produce the required instruments to the faculty's complete satisfaction, it was still far from certain that he could be employed as the university mechanic since no funds were available at that time for such a position. There is no record of Carl Zeiss having sent a reply. His dire financial situation would probably have rendered him incapable of producing any sample instruments.[12]

This episode is significant, not only in biographical terms, but also for the evolution of ZEISS as a company. If Zeiss had relocated his workshop to northern Prussia, the subsequent collaboration between Zeiss and Abbe would most likely never have happened – and the company would not have enjoyed such a meteoric rise in its fortunes.

The entry regarding the nuptials of Carl Zeiss and Bertha Schatter in 1849 from the marriage register in Neunhofen.

Zeiss stays put, but starts afresh

We actually know relatively little about Carl Zeiss's personal life. In 1853 he remarried, this time to Ottilie Trinkler (1819–1897), the daughter of a clergyman from the town of Triptis in eastern Thuringia, who was distantly related to Zeiss's first wife.[13] Although Ottilie was a ninth-generation descendant of Martin Luther, Zeiss never made much of this fact, unlike subsequent biographical researchers who saw this distant family relationship as an indication of the company founder's Protestant beliefs.[14] Carl Zeiss himself subsequently referred to both his wives (affectionately) as "spiritually very much country folk."[15] This is a striking comment, since Jena itself was something of a backwater at that time, with neither railway connection nor any industry of note. The bond between Zeiss and his first wife's family remained firm, not only due to the period they spent caring for his son Roderich, but also as a result of the marriage of Zeiss's sister Pauline

Ottilie Zeiss, née Trinkler, second wife of Carl Zeiss.

to the widowed Carl Gustav Schatter. Thus, Zeiss's first father-in-law simultaneously became his brother-in-law. Zeiss's marriage to Ottilie produced three children: Karl Otto (1854–1925), Hedwig (1856–1935) and Sidonie (1861–1920). The relationship between Zeiss's first son Roderich (1850–1919) and his stepmother was never very affectionate, and he was soon sent to a grammar school in Eisenach.[16]

At around this time, when Zeiss was becoming acquainted with the woman who would become his second wife, he began to take a serious look at how to strategically develop his business. Zeiss was not merely suffering from a temporary sales crunch. He believed that microscope manufacturing – and therefore his company – had reached an impasse. In 1855 he wrote a long letter to his friend K. O. Beck in Moscow in which he reflected soberly on the years that had passed since he founded the company.

"Before I had even finished setting up the business, Körner passed away in February 1847. Since then I have primarily been engaged in making the small kinds of 'simple' microscopes. I was able to produce them slightly more cheaply than Körner and much more cheaply than Plössl [= Simon Plössl in Vienna], [...] so I have always done much of my business

Otto Zeiss, the second son of Carl Zeiss, circa 1869; photograph from the collection of Kathrin Siebert.

Sidonie, the youngest daughter of Carl and Ottilie Zeiss.

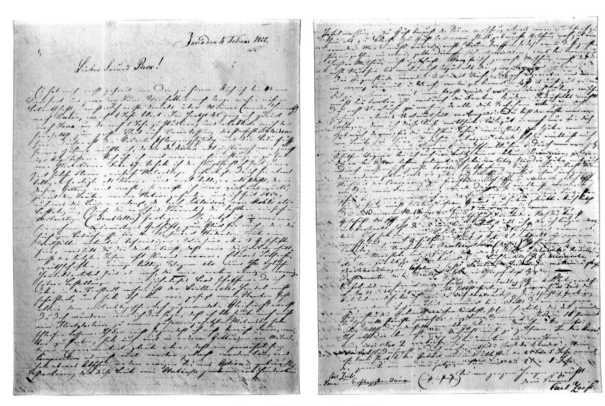

First and last pages from the letter to his friend K. O. Beck (most likely Karl Otto) dated 1855.

abroad. People also liked my microscope stands and since 1852 Schleiden, von Mohl etc. have confirmed – also in regard to lenses – […] that I make the best simple microscopes (doublets).

Now some two-thirds of my income comes from export business, fortunately for me since the local requirements for the 3½ hundred students could be easily met by just one workshop. Yet the reality is that there are three businesses here, so, as you can imagine, I have been through some testing times in recent years. It is with items such as a blowpipe, some cheap carts etc. and optical merchandise that I have managed to earn a crust, together with some all too infrequent one-off orders. The fact that I do not live centrally means that there is limited business with spectacles and such like, which is why, as I stated before, I have applied virtually all of my efforts to the realm of microscopes.

All the same, you will be surprised to hear that I have not yet had a flint lens ground for a compound microscope. Do not think that I would not be ready to make further endeavors in this area – indeed I have experimented with many other classes of microscope – yet I believe that the usual compositum cannot take us much further, and I have something of an aversion to this relentless rigmarole of trial and error that is so common among us lensmakers, with people such as Oberhäuser attempting to make one good objective lens from among hundreds of lenses."

This letter makes it clear that there were deep-rooted reasons behind stagnating sales. Local demand for spectacles, simple microscopes and precision mechanical instruments was far too limited to enable Zeiss to support his growing family. At that time Zeiss was already reliant on the export market for two thirds

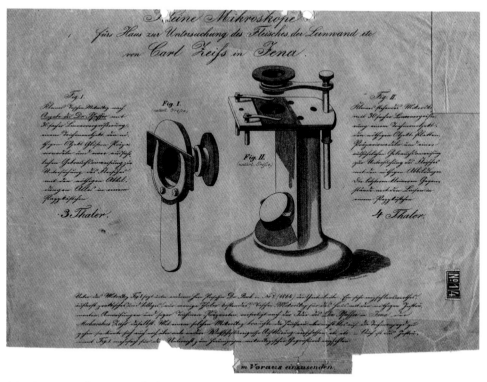

Advertisement: small microscopes for domestic use to examine meat, canvas, etc. produced by Carl Zeiss Jena circa 1865.

of his sales. The German conglomerate of separate states and kingdoms meant that Jena was remarkably close to many of the borders of Saxe-Weimar-Eisenach, a fact that virtually obliged Zeiss to specialize in technologically sophisticated products which were not available locally to customers from outside Jena. Zeiss had decided to focus primarily on building microscopes, and by now he had made something of a name for himself in the field of simple, single-lens microscopes, which today resemble little more than a powerful magnifying glass with a stand. Yet it was already clear that the future lay in compound microscopes, which consisted of both ocular and objective lenses. A well-known exponent of these instruments was Georg Oberhäuser (1798–1868) from Ansbach, who Zeiss mentions in his letter. Together with his employee and eventual successor Friedrich Edmund Hartnack (1826–1891), Oberhäuser ran one of

Europe's most important optical workshops in Paris. Although Oberhäuser's compound microscopes were considered to be some of the best in their field, they – and indeed all the competing products – were based on a highly flawed method.

A science, not a craft

None of the microscope makers at that time had a clear understanding of the mathematical and physical principles which lay behind the interaction of the different lens systems in a microscope. Moreover, there was huge variation in the quality of the available optical glass, so the standard method was simply to keep combining different pairs of lenses until they formed a usable optical system, essentially relying on a process of trial and error. As a result, building compound microscopes at that time was something of an art which depended heavily on the individual talent and experience of the lensmaker concerned. This was one of the key reasons why none of the leading microscope providers had considered tackling the pervasive supply shortages in the microscope business by setting up a factory-style manufacturing operation.

As a newcomer to the market, however, Carl Zeiss could not afford to spend decades building up a level of experience that would enable him to catch up with the competition. What is more, the principle of trial and error without any guaranteed result was contrary to his very nature, as indicated by the letter to K. O. Beck cited above. He therefore decided to forge his own path which could somewhat anach-ronistically be described as upholding Communist values. His idea was simple: if he could develop a precise mathematical model of how a microscope works and, at the same time, produce lenses with precisely defined optical parameters at a consistent level of quality, then he would transform the art of microscope making into a manufacturing process based on scientific principles. The economic benefits could hardly be overstated: a company operating on a scientific basis would be able to mass-produce powerful lenses at a consistent quality and with far less wastage.

This idea was by no means new. Joseph von Fraun-hofer had already laid the foundations for creat-ing a modern optics industry decades earlier. From 1806 onwards Fraunhofer dedicated himself to the

challenge of producing improved optical glass at the Mathematical Mechanical Institute in Munich. His research into spectral lines and the diffraction of light enabled Fraunhofer to measure the properties of optical lenses with hitherto unsurpassed precision, paving the way for the industrial manufacturing of precision lenses. Fraunhofer also made spectacular progress in telescope design, basing his models on exact mathematical calculations instead of relying on experience like his contemporaries. The two identical telescopes for the observatories in Tartu (Estonia) and

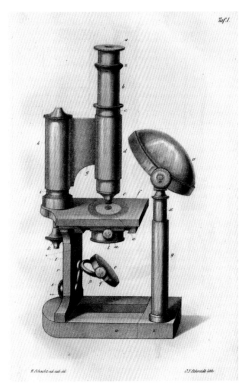

Large microscope produced by Oberhäuser, sketched by Hermann Schacht, from his book "Das Mikroskop," 1855.

Joseph von Fraunhofer (second from left) demonstrates the spectrometer to Joseph von Utzschneider, Georg Reichenbach and Georg Merz; painting by Richard Wimmer, 1900.

Berlin, which came into operation in 1824 and 1829, show just how much he was able to achieve with this approach. For astronomers at the time, these telescopes represented the pinnacle of what was technologically feasible – and they enabled revolutionary discoveries such as that of the planet Neptune in 1846. Essentially, Fraunhofer made the advances in the field of telescopes in the mid-1820s which Carl Zeiss would seek for microscopy some 30 years later. So why wasn't Zeiss able to use Fraunhofer's work as a basis for making microscopes in Jena? After all, Fraunhofer was also known as a manufacturer of very powerful achromatic microscope lenses.[17]

Fraunhofer's hidden legacy

Unfortunately it was extraordinarily difficult and in many cases even impossible for Zeiss to draw on Fraunhofer's insights and findings. The pioneer of science-based lensmaking spent his career in a field in which trade secrets were jealously guarded. Fraunhofer's business partners were so shortsighted that they even destroyed important documents written by Fraunhofer following the sudden death of the brilliant researcher in 1826. Much of his practical know-how was therefore denied to the following generation. One example is the method of checking lenses using a trial lens. This method was 'discovered' by August Löber, Zeiss's first assistant, in 1861, even though

Fraunhofer had actually applied it years earlier.[18] There is good reason to doubt that Carl Zeiss would have been familiar with any of Fraunhofer's theoretical work. Fraunhofer enjoyed less of a reputation during his lifetime than he does today, and Zeiss was not a systematically trained scientist, but rather a practitioner possessed of some theoretical knowledge. Zeiss simply suspected that a series of unknown mathematical and physical obstacles were stubbornly entrenched in his field of work. It is a testament to his visionary power that he was willing to make a risky investment in basic research at a time of such economic hardship.

The search for a scientific basis

Zeiss began searching for a qualified scientist to work with him as a partner, and his first choice was the mathematician Friedrich Wilhelm Barfuss[19] (1809–1854). Their collaboration was in full swing by 1852 at the latest, and it seems the two men became quite close, with Zeiss describing Barfuss as his "friend" upon the latter's death.[20] This can be explained by the similar paths their lives had followed up to this point. Both came from families of craftsmen from the Grand Duchy of Saxe-Weimar-Eisenach (Barfuss: stocking makers; Zeiss: wood turners). They also both attended grammar school in Weimar, though the age gap of seven-and-a-half years means that they are unlikely to have become acquainted at this time. It is therefore reasonable to assume that they did not meet until later in life.[21] Barfuss, who obtained his doctorate in Jena in 1838, was working as a private tutor in

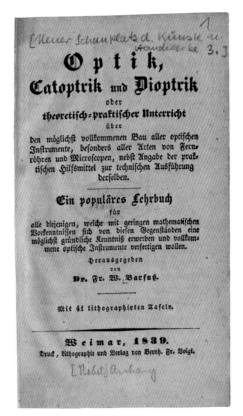

Cover of a textbook by Friedrich Wilhelm Barfuss, 1839.

Zeiss's native town of Weimar while also accepting work from various clients on a self-employed basis. These included Zeiss's teacher, the mechanic Friedrich Körner, who sought out Barfuss's assistance in calculating optical systems.

A collaboration with Barfuss seemed promising because he had already published a brief treatise on compound microscopes in 1846. This included a description of how to correct spherical aberration – an optical effect inevitably present in all simple microscopes – by dividing the optical system into an ocular lens and an objective lens, whereupon Barfuss noted that "this method has been tested through careful calculations taking into account all the circumstances."[22] Yet ultimately he could not deliver on this promise. When Barfuss died suddenly in 1854, Zeiss was still nowhere near achieving his much-anticipated breakthrough in compound microscopes. It would nevertheless be inaccurate to class Barfuss as incompetent, as so many Zeiss biographers have done based on a scathing verdict by Ernst Abbe.[23] The Barfuss-Zeiss collaboration did produce some useful results, for example triplets – i.e. simple microscopes made using three lenses – designed on the basis of mathematical calculations, which were sold until at least 1868.

Yet whatever success Zeiss had in refining the lenses he had been selling since 1847, it was abundantly clear that if he wished to become more than just a lensmaker who produced instruments for university education and simple dissection tasks, he would have to accept that the principle of the simple microscope had long since reached its limits. Zeiss's mentor Matthias Jakob Schleiden had realized this as early as 1842:

"The question remains as to whether the simple microscope or the compound microscope is more advantageous for scientific experiments. I would most definitely choose the latter [...]. [...] If one merely compares the observations made over the past 20 years, it is impossible to deny that – with the exception of Robert Brown's discoveries [...] – all the observations which have promoted the cause of science have been made exclusively with the compound microscope."[24]

The first compound microscope

The crucial decision on the future of microscope development had therefore already been settled by the scientific community. Zeiss was consequently left with no choice but to respond by applying himself to the far more complex realm of compound microscopes. In 1857 he sold the first of these instruments. It was built by using the existing two-lens optics system of a simple microscope as objective lenses, combining this with an ocular lens, and connecting the two elements via a tube. The prototype was purchased by the botanist Hermann Schacht, who had also acquired the first simple ZEISS microscope 10 years before. The company officially included the compound microscope in its product range from 1858 onwards. Its description in the price list ran as follows:

"Small lens barrel consisting of a tube with a collecting lens and two ocular lenses with a contrivance for connecting the

One of the earliest compound microscopes from 1862, from the collection of Timo Mappes; photographed by Manfred Stich.

tube with the doublets and stands no. 1 through 5 and, using the doublets as an objective lens, achieving two higher magnifications based on the principle of the compound microscope. This allows the 120x magnification of the simple microscope to produce a 300x and a 600x magnification. The whole device is presented in a hardwood case, including a concave mirror which must be fabricated to fit on the stand's flat mirror in order to produce brighter illumination."

In that same year Zeiss published a short article[25] in the *Annalen der Physik und Chemie* in which he described a phenomenon he had observed during his experiments with a wide variety of microscopes. Under poor illumination of an object, the image shifted sideways whenever the lens barrel was moved up or down. Zeiss had no idea what was causing this effect, though he had ruled out the possibility of flawed mechanical construction. Although this represents just one brief episode, it shows the position Zeiss was in some 12 years after founding his company. Although his designs could hardly be improved in terms of their mechanical construction and quality, Zeiss was still faced with phenomena which, as a practical lens-maker, he could merely diagnose but not explain. His attempt to "buy in" scientific knowledge to further his development work may have met with only limited success in the case of Barfuss, but the idea in itself was both correct and critical to his survival.

Yet the dire economic situation of Zeiss's workshop remained unchanged. Schleiden praised the new instruments, yet actual sales were negligible, with only 17 compound microscopes being built between 1857 and 1859. In June 1859, the number of wage-earning employees at ZEISS fell from three to two, as recorded in the 'Manual' kept by its owner. Zeiss was at a crossroads. If he wanted to steer his company out of the crisis, he would have to take the bull by the horns and position himself firmly at the cutting edge of scientific and technical development instead of simply reacting to developments as they unfolded.

"Part of Our DNA" – Carl Zeiss and ZEISS in 2016:

An Interview with Prof. Dr. Michael Kaschke, President and CEO of Carl Zeiss AG

What do you associate with the man that was Carl Zeiss?

In my opinion, our company founder represents certain virtues that still make our company what it is today: a dedication to exceptional quality, a healthy thirst for experimentation, the willingness to take entrepreneurial risks and the perseverance you need to pursue a promising technological endeavor from the initial idea to a viable product.

In the mid-19th century, there were a lot of experts who simply did not believe that we would ever be able to produce microscopes on the basis of scientific calculations alone without having to endure

lengthy trial and error processes. But Carl Zeiss was not one to give up so easily – which is what led him to lay the foundation for what would later become a successful company. It is precisely this belief – that limits are surmountable – that is indispensable to us today. During my postgraduate studies in Jena in 1988, Ernst Abbe's resolution limit for optical systems was hailed as a kind of ultimate natural law. At the beginning of the 21st century, it was ZEISS that surpassed this limit by constructing special fluorescence microscopes. So if there is one thing we can learn from Carl Zeiss, it is that in our industry we can push the limits of possibility time and again through an unwavering drive to develop.

**Is it not in fact Ernst Abbe who should be cred-
ited with laying the scientific foundations that
led to the construction of the microscope? Is
it possible that, as an entrepreneur, Carl Zeiss
simply had the good fortune to work alongside
such a brilliant scientist?**

The relationship between Abbe and Zeiss is often
boiled down to a juxtaposition between scientist and
entrepreneur. I think it's important to make people
aware of Abbe's business acumen and likewise about
the pronounced fundamental scientific understanding that Carl Zeiss possessed. Zeiss's correspondence
with numerous renowned scientists demonstrates
that he wanted to know how the instruments he
built were used. This ability to interact with the
scientific community ultimately made his business
venture a success – and this still rings true today. I've
just come back from Shanghai, where we opened
our new Chinese headquarters a few days ago. What
really made an impression on me were just how
many respected scientists from a host of different
disciplines attended the event. It may seem somewhat far-fetched to you, but when the scientific community takes such a great interest in our company it
is a reflection of all that we have inherited from Carl
Zeiss and Ernst Abbe. This symbiosis of business and
science is part of our DNA.

**Carl Zeiss worked closely with Matthias Jakob
Schleiden, a botanist in Jena, to develop his first
microscopes. Schleiden was also one of the first
ZEISS customers …**

In the company's infancy, a high level of customer
focus certainly went a long way, but Carl Zeiss soon
had to realize that this approach could also be a dead
end. As the 1860s came to a close, this is exactly
what happened: owing to the high level of customi-
zation of his products, Zeiss had to juggle a complex
portfolio with a great many niche products without
even one of his microscopes achieving anything that
came close to the quality of the best products his
competitors were offering. And this in spite of the
fact that he always took his customers' suggestions
for improvement very much to heart. The challenge
was to take the market's reaction and use it to drive
the design process. With Abbe's help, Zeiss put
microscope construction on a scientific basis. In so
doing, he laid the foundations for a product range
that met his customers' needs better than they had
ever thought possible.

**So you're saying he understood his customers
better than they understood themselves?**

He understood the ambitions of his customers and
then developed the technology that most closely
satisfied them. I see this as true customer focus: it's
not a question of asking users about the minutiae,
it's about finding out what it is they want to achieve
using our instruments and machines. I think this is
where Carl Zeiss underwent a development process:
at the beginning it may well have been his aim to
meet every last customer need. But the transition to
volume production forced him to design his range
of models in such a way that he could fully utilize an
efficient production process – while maintaining a
consistent level of quality.

"A consistent level of quality" sounds a little too modest a way of describing ZEISS. Is the brand not a symbol of quality that makes no compromises?

There is simply no absolute measurement for quality. A product can only make no compromises if considered in relation to a particular application. This was something Carl Zeiss was very much aware of – not least of all because of the close contact he had to users themselves. They did not have unlimited funds for their facilities and numerous microscopes were also used to teach students. The main demands were ease of use and good image quality for the price. At the same time, quality at ZEISS – both in the past and the present – does of course play a major role: we want to build the best instruments on the market within the scope determined by a specific application.

As a businessman, Carl Zeiss took a great risk by deciding to build a microscope using scientific calculations. Economically viable results could only be expected over the long term and were

by no means guaranteed …

Right, but it's important to understand that this courageous willingness to take a calculated risk was one of the keys to his success. As a company fueled by innovation, we constantly face the challenge of breaking new scientific ground. Many projects are fraught with uncertainty in terms of their economic viability. Such projects can only be successful if those working on them see eye to eye. This is yet another area in which Carl Zeiss was particularly adept: he was able to create a corporate ethos that appealed to the brightest minds of his time. And the involvement of a scientific and technical virtuoso like Ernst Abbe gave Zeiss the confidence he needed to take his risks.

There's another aspect I think is important: Carl Zeiss had the foresight to partner with trustworthy individuals when he knew he had reached his limits. This ability of being open to new ideas from outside was something that occasionally fell by the wayside at Carl Zeiss. In the 1970s and 1980s it was a question of honor to be able to do everything on one's own. For Carl Zeiss in East Germany, this was an attitude born of necessity; in West Germany, it was part of a positive self-perception. In the two decades that followed the reunification of the two companies, we bore witness to a paradigm shift. Today, our strategy bears more of the hallmarks of our founder Carl Zeiss, who never ruled out lasting partnerships. Recently, the value of cross-company development activities was illustrated by our partnership with ASML in the Netherlands with the production of lithography systems for the semiconductor industry. The breakthrough came in 2004 with a system based on the principle of immersion lithography. I am convinced that, if working alone, neither ASML nor ZEISS would have been in a position to bring a development of this magnitude to market with such speed.

ASML was founded in 1984. ZEISS can already look back on 170 years of corporate history. But is this tradition not in fact a burden? Would there be something missing if we stopped thinking about a founder such as Carl Zeiss?
You know, the fascinating thing is that someone born almost 200 years ago stood for values that are still relevant today. There's no doubt that it is also possible to create leading-edge technology without a long tradition to back it up. However, being able to return to the origins of ZEISS can help us see what it is we need to preserve. Let's assume for a moment that Carl Zeiss could travel through time and see his company as it is today. I think he would certainly see a lot of himself in it.

Prof. Dr. Michael Kaschke (*1957 in Greiz) studied physics at the University of Jena where he wrote his dissertation, earning the titles Dr. rer. nat. and Dr. sc. nat. He worked at various research institutes before joining ZEISS in 1992. He has been a Member of the Executive Board since 2000 and President and CEO since 2011.

Flourishing on a Fragile Foundation:
A Booming Business and
Social Recognition (1859–1866)

Initially, Carl Zeiss continued to face major technological hurdles inherent in manufacturing precision optics. Nevertheless, he succeeded in making a name for himself as an entrepreneur both in business and in society. Evidence of this came in early 1858 when he purchased the property at Johannisplatz 9, which abutted Jena's historical center. The building, which Zeiss procured using the 500 thalers borrowed from his sisters Emilie and Pauline[26], was spacious enough to accommodate the subsequent expansion of his business. In addition to the workshop in the annex, the retail outlet on the ground floor and the private quarters on the first floor where the Zeiss family dwelled were large enough for them to rent a sizable apartment to the gynecologist Bernhard Sigmund Schultze in January 1859. In 1860 Zeiss also acquired an adjacent building in the courtyard to expand his workshop.

At first, microscope sales continued to stagnate, so most of Zeiss's turnover came from other products. In addition to repairs and commissions for university research departments, the sale of merchandise and spectacles to private individuals played an important role, as the following advertisement, published in December 1860, demonstrates:

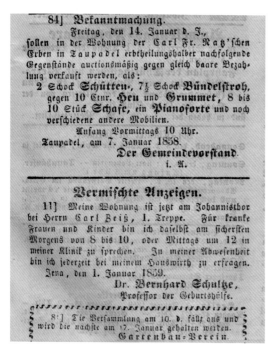

Rental advertisement from Dr. Bernhard Schultze in the "Blätter von der Saale" dated 6 January 1859.

Johannisplatz in Jena, circa 1865; the second house from the left was the location of the third Zeiss workshop.

Advertisement in the "Blätter von der Saale" newspaper dated 6 December 1860.

"What makes an excellent Christmas gift? Telescopes, single and double opera glasses, lorgnettes, pinces-nez and eyeglasses made from different materials, spectacle frames and pinces-nez in gold, exquisite Parisian and handmade spectacle lenses, loupes, microscopes, barometers, thermometers, drawing sets, letter scales, etc.
Carl Zeiss."

A glance at Zeiss's accounts (his 'Manual'), which he managed single-handedly, shows us the value of the advertised items. During the month in which the quoted advertisement appeared in *Blätter von der Saale*, Zeiss devoted much of his time to ophthalmic work, yet he also found the time to carry out a great many repairs. In the 'Manual,' frequent mention is also made of thermometers, different drawing sets (i.e. exceptional drawing instruments for scientific and technical applications) and telescopes.

A master of weights and measures

Zeiss's skills as a lensmaker and mechanic soon garnered him official recognition. In July 1858, the Jena-based entrepreneur was appointed by the Grand Ducal Office of Weights and Measures in Weimar as Deputy Master of Weights and Measures and tasked with overseeing weights and measures in and around the city of Jena. As local commerce and the calculation of possible levies, such as on agricultural products, depended on the 'standard measures' Zeiss was to prescribe in his new role, this position came with significant responsibility. Zeiss had already served in this position since 1852 under the tutelage of the local shoemaker Vater. For many years Zeiss had resisted Vater's condescending in particular because he feared that an ordinary craftsman might accuse him of errors in front of his own employees. Initially, the local council in Jena turned a deaf ear to Zeiss's entreaties and in 1856 Zeiss declared with some degree of resignation that he would not continue to hold this position given the prevailing circumstances. For Zeiss, it was:

"abhorrent [...] to act as a mere mute in the Inspection Office and to bow down to a man such as the shoemaker Vater who revels in controlling all things with such pedantic rigor, who barely understands my business and to whom I am nothing more than a puppet."[27]

After around two years, in which there was great uncertainty about how this position would be filled at the Jena Office of Weights and Measures, Zeiss's

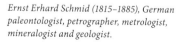

Announcement from the "Blätter von der Saale" newspaper dated 12 July 1858.

entreaties in 1858 were acknowledged by a higher authority and he was permitted to guarantee the correctness of weights and measures for all tradesmen in Jena solely on the basis of his own inspection.[28]

A university mechanic

On 7 June 1860, Zeiss's colleague Johann Friedrich Braunau (1810–1860) passed away. He had held the position of University Mechanic in Jena since the death of Friedrich Körner. Approximately six weeks later, Zeiss wrote a letter to the prorector of the university and the university senate in which he requested that he be appointed to the now vacant position. He included five letters of recommendation from

professors, all of whom vouched for his suitability both professionally and personally. It was clear that Zeiss had spent the last few years building up a base of satisfied customers from among Jena's scientific community. The geologist and paleontologist Ernst Erhard Schmid wrote the following:

"I have known Mr. Zeiss for a great many years. He possesses the requisite theoretical and practical knowledge, is diligent in everything he does, is highly ambitious, always agreeable and is unquestionably an honorable gentleman. His workshop is well-equipped for all manner of high-precision mechanical work. His business is highly spoken of and recommended far and wide because he assiduously and skillfully fulfills his orders and does so punctually and economically. I [...] would have no qualms in offering the vacant position to him as well-deserved recognition of his many years of service to our university and the fruits his services have borne."[29]

As a result of such a sterling show of support from many professors, Zeiss's request also found favor with the prorector of the university at the time, Heinrich Luden, and the senate of the Jena *Alma Mater*. Nevertheless, there was disagreement as to whether Zeiss should simply be granted

Ernst Erhard Schmid (1815–1885), German paleontologist, petrographer, metrologist, mineralogist and geologist.

the title of university mechanic or whether – unlike with Braunau – the creation of a new position as a lecturer was in order. For Zeiss, being granted permission to teach was an interesting development as he assumed that he would receive a generous tax exemption based on a law that dated back to 1851. It was very good news for the company when, on 10 September 1860, the government of the Grand Duchy approved the university's proposal regarding his appointment as a university mechanic. On 25 May 1861, the Grand Ducal Saxon Tax Collection Office in Jena also confirmed that Zeiss would henceforth be exempt from paying income tax due to his position as a university lecturer – a distinct competitive advantage for the small workshop. Less than two weeks later, however, the higher-ranking authority in the state capital of Weimar objected to this classification and declared that the exemption would apply only to income received from the university. The earnings "from the workshop founded in the city of Jena and the appertaining shop"[30] were all subject to taxation. But Zeiss did not give up so easily. He wrote a letter to the university senate in which he requested that they pursue the issue of tax exemption on his behalf. He also expounded on the details of his present working situation. Even if we cannot begin to assess the information – and it is likely that Zeiss slightly exaggerated the facts in order to give weight to his argument – this self-portrayal is revealing. For example, it contains the statement that the Optical Workshop in Jena filled orders primarily for members of the scientific community. Zeiss asserted "that any articles (spectacles and so forth) I sell to non-academics account, at best, for no more than 1/20 of my

The main building at the University of Jena, the so-called 'Wucherei' (1858–1908).

total sales." He also felt "a moral duty to accept every single order placed by an academic (even if they yield no profit for me)."[31] Zeiss went on to cite precedent for a considerable tax exemption, drawing on the cases of the photographer and dance teacher hired by the university – two positions which were clearly less important than that of a learned mechanic.

Zeiss embroiled in tax dispute

Thus presented with Zeiss's arguments, the prorector saw no other solution than to seek legal counsel. It was found that the decision not to grant a tax exemption was justified in Zeiss's case, as income earned through self-employment was not covered by the law. And so it was that in July 1861 Zeiss received word that he would not be able to count on the university's support in his dispute with the tax authorities. It is unlikely that the newly appointed university mechanic enjoyed any commercial advantages as a result of his position; after all, he had been receiving commissions from the university in Jena prior to his appointment. Nevertheless, he was now privileged enough, or rather obliged, to teach at the university. As is documented in the lecture schedule for the university in Jena, Zeiss gave students the opportunity to try their hand at "producing physical and optical devices" every semester until his passing in 1888.[32] We do not know whether the students took advantage of this offer and how much time Zeiss actually invested in his teaching.

We can say for certain that Zeiss saw himself first and foremost as an optics entrepreneur and not as a service provider for the university – despite his attempts to convey a different impression in his struggle to gain tax benefits. Zeiss's attempt to direct his business toward the free market is evidenced by decisions such as his participation in the General Thuringian Trade Exhibition in the summer of 1861, which saw 1,300 exhibitors make their way to Weimar. Zeiss was one of the 28 winners and received a top honorary award with an embossed gold medal. The reason: Zeiss was being honored "for one of the most sublime microscopes ever produced in Germany that allowed for admirable flat, razor-sharp and bright images with all lens systems."[33] Further proof of his growing reputation came in 1863 when Carl Zeiss was named Grand Ducal Mechanic of the Court.

Zeiss had now been elevated to the same level as his teacher Friedrich Körner, who had been Mechanic of the Court until his death in 1847. Zeiss's new title appeared to change very little in practice. Even so, his prestigious new title was used in the university course catalog and in newspaper advertisements, where Zeiss was often referred to as Mechanic of the Court – and not as a university mechanic.

Even though Zeiss garnered an ever-greater reputation as a manufacturer of optical devices in other regions, he still relied on local business as his primary source of income. This theory is supported, for example, by two (identical) advertisements[34] that were published in *Blätter von der Saale* in August 1861.

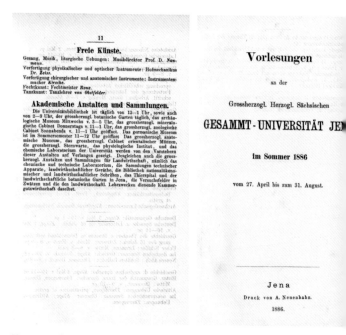

University of Jena lecture schedule, August 1886.

The exhibition building of the 2nd General Thuringian Trade Exhibition in Weimar, 1861, pictured in the weekly magazine "Illustrierte Zeitung" no. 936, on 8 June 1861.

Advertisement from the "Blätter von der Saale" newspaper dated 13 August 1861.

These reveal that the Berlin-based company Julius Imme & Co. had Zeiss as the sole sales representative of a "volta-electric metal brush" in the Jena area. This device purportedly cured various ailments such as rheumatism and gout by applying an electrical current to the skin. This technology, which was just coming into vogue at the time, was far removed from Zeiss's core business.

A stint as a local politician

In November 1861, Zeiss ran for a position in the city council for the first time but was not elected. It is unclear whether the university mechanic was pursuing any particular political aims or whether he was primarily interested in having his say in important

local issues that concerned him as an entrepreneur. Whatever his reasons, Zeiss ran once more in November 1863. This time, he was the seventh of twelve candidates to be elected to the city parliament, having received 197 votes out of a possible 363. Over half of the citizens who voted in Jena wished to see Zeiss become a member of parliament. At the same time, he was named *Armenpfleger* in his district, thereby becoming a volunteer in charge of donations and funds for local welfare.

He remained politically active until at least 1867, when he decided against running for Jena the municipal council again. Very little is known about Zeiss's political goals. There is only one instance we know of today in which he pursued a specific initiative: in January 1867, he signed an open letter with 21 other

Open letter against the German People's Party from the "Blätter von der Saale" newspaper dated 11 January 1867.

citizens to express opposition to the aspirations of the newly formed German People's Party to make itself a representative of the entire nation through its name. "We [...] believe," read the text that appeared in the *Jenaische Zeitung* "we have a heart that beats for the nation just like they do and [...] we know that we, belonging to the nation, share [...] the same interests as the majority of them."[35] Based on the left-wing lib- eral aims of the German People's Party, which stood, in particular, for rapprochement with Austria and against the dominance of the Kingdom of Prussia, we can hazard a guess as to the political leanings of Zeiss. By signing the protest, Zeiss identified himself as someone who tended to ally with the national liberals who hoped for a unification of the German empire under Prussian rule. The letter also stresses the fact that the movement garnered a majority among the "people" (i.e. from the male citizens eligible to vote) and shows that Zeiss would not have held any extreme political views.

A modular portfolio

Let us go back a few years to the beginning of the 1860s, when Carl Zeiss enhanced the inner work- ings of his microscopes. Even though the optical and

Market square in Jena circa 1865; by A. Meysel.

manufacturing hurdles persisted – each instrument still had to be assembled in a laborious process of trial and error – Zeiss nevertheless made remarkable headway in terms of the mechanical design. He had previously used the lenses and stands from simple microscopes to construct compound microscopes. But in August 1861 he began offering his own range of five stands specially developed for compound microscopes. The largest of these stands, as detailed in the price list, was the 'large horseshoe stand.' Zeiss took his inspiration from the renowned lensmaker Georg Oberhäuser (Paris).[36]

One particular feature of this model was the cylindrical diaphragms used to control the aperture. These were not integrated into the stage from below but from the side using a slide mechanism. This was a more sophisticated design and placed greater demands on precision during manufacture, but it prevented the rotating mirrors, integrated below, from being shifted inadvertently when changing the diaphragm. This detail was a real advantage for those who worked with a microscope on a daily basis. A photo of 45-year-old Carl Zeiss from 1861 shows the aforementioned horseshoe stand as an allusion to his profession. We can also clearly see how proud he was of his product. The horseshoe form would characterize the design of the larger ZEISS microscopes for decades to come.

In the meantime, Zeiss's portfolio had grown substantially and now featured various stands, lenses, eyepieces and accessories. Customers who placed an order for a particular instrument could select

Horseshoe stand from 1874, from the collection of Timo Mappes; photographed by Manfred Stich.

compound microscopes and I was fortunate enough to inspect several models, of different sizes. His compound microscopes are distinguished by their incredible speed and by the flatness and sharpness of the image for all sizes, and the microscopes are furnished with stands that are as functional as they are meticulously executed. [...] Even the smallest stand, no. 4, [...] is suitable for the most high-powered objective lens systems, which distinguishes the smaller

Carl Zeiss aged 45, with horseshoe stand.

from among a wide range of different combinations. Zeiss charged for each part separately. Customers paid only for the equipment they required. The high degree of modularity was also warmly welcomed in professional circles, as evidenced by the following statement made by the aforementioned botanist Hermann Schacht in *Das Mikroskop und seine Anwendung*, a school textbook from 1862:

"Carl Zeiss from Jena, whose simple microscopes have been known far and wide for quite some time, has recently begun manufacturing truly excellent

Catalog and price list for microscopes and accessories, 1858.

Letter of recommendation for Zeiss microscopes written by Prof. Schleiden.

microscopes offered by Mr. Zeiss from those manufactured by other lensmakers. The weaker objective lens systems – A, B, and C – offer exceptional speed […]. […] Zeiss leaves it entirely up to the customer to select objective lens systems and ocular lenses and charges for the individual parts." [37]

All one needs to do is look at Zeiss's sales numbers to see that his strategy was a great success. Just 15 microscopes were sold in 1860 – the lowest sales figure since the workshop had been established – whereas 192 systems were sold by 1866, most of them compound microscopes. The reputation of the Optical Works was growing, but this also resulted in an increasing number of supply shortages. In 1847 the scientist Schleiden had praised the fact that "Mr. Zeiss always has so many instruments available for purchase that all orders he receives can be fulfilled almost immediately." [38] 15 years later, the

situation was very different. Zeiss frequently found himself having to ask that his customers be patient and it was plain to see that he did not like being behind schedule. This letter, dated August 1862 and written to schools inspector Dr. Eberhardt in Coburg, demonstrates how Zeiss, in a precarious situation, attempted to strike a balance between offering high quality and prompt service:

"Allow me to be frank: for some 5 weeks now your highly honorable commission was the oldest of all my orders from afar; this is due to there being an older one originating closer to home. Indeed, my intention was to send you the articles several weeks ago, however I hoped first to have a larger selection of systems available and have been waiting until some more […] were ready. […] In regards to the magnifications detailed in the catalog: I hasten to add that the figures have been rounded and,

77

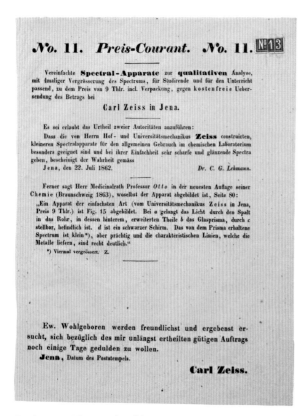

Catalog page with note on late deliveries, 1862.

cess. [...] You will receive the best product that I can currently offer.
My most sincere regards and humility,
Carl Zeiss

Please examine the lenses with a loupe before the initial inspection." [39]

Concerns about the future

But it was not merely rising demand that exerted pressure on Zeiss. Commercial success also heightened the risk of his business secrets finding their way into the hands of his competitors. Journeymen whom Zeiss temporarily employed could potentially pass on detailed key knowledge and the tricks of the trade used in manufacturing optical instruments. It was also conceivable that competitors would resort to bribery to acquire company secrets. Zeiss made every effort to at least establish legal and moral defenses. New employees were obliged to take an oath under penalty of "temporal and perpetual punishment" should they break their vow of secrecy. The declaration from the optician Fritz Müller has been preserved. In October 1866 he pledged to observe the following:

"I, Fritz Müller, do solemnly swear before almighty God as our all-knowing and all-powerful creator to uphold this oath: that I, as a current and future maker of lens systems and such like under the employ of Mister Carl Zeiss, Court and University Mechanic, shall now and forever more closely guard

as such, are not to be taken as wholly accurate. I hereby inform you that each system has, ceteris paribus, a particular magnification that no other system has because the radii and distances and the optical characteristics of the lens have consciously not been made to be identical. The systems that afford higher definition must therefore [...] be mounted twice as I must always first experiment [...]. [...] Allow me to express once again how grateful I am for your forbearance and patience during this lengthy pro-

Fritz Müller (1847–1919; 1861 to 1913 at Carl Zeiss): in 1864 as an apprentice and in 1913 as a foreman.

to assume that the workshop in Jena had already ascended into the upper echelons in this key area for the future development of microscope optics. And yet Zeiss still looked to the future with caution: in the eyes of his peers, the devices he produced were certainly admired, but esteemed researchers such as the botanist Leopold Dippel (1827–1914) also stressed that the microscopes produced by Carl Zeiss came nowhere near approximating the immersion optics from the workshop of Edmund Hartnack (1826–1891) in Paris.[41] Carl Zeiss was surely very much aware that he was once again at risk of failing to keep abreast of technological progress. He

all the numbers, measures and the methods of lensmaking developed by this company to which I have access and shall never, for any purpose whatsoever, write down any secrets from the company for personal use nor use them for other purposes other than for those that benefit the company bearing the Zeiss name nor in any way that impairs this pledge of secrecy; I swear to be a faithful worker to Mister Zeiss, Court and University Mechanic, and always to set a fine example for younger workers who enter the company from this moment on by virtue of my truth, loyalty and punctuality; so help me God and His Holy Word through Jesus Christ my Savior and Redeemer, Amen."[40]

Zeiss expressly prohibited his assistants from compiling notes on manufacturing techniques and taking them off the premises; his main concern was to safeguard the "methods of lensmaking" and the measures and optical parameters. It is reasonable

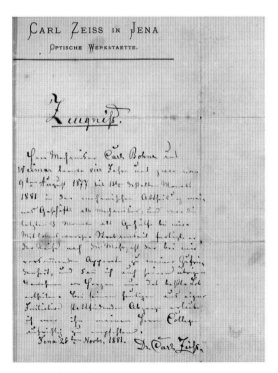

Letter of reference for the mechanic Carl Bohne, 1881.

The largest microscope from the workshop established by Georg Oberhäuser, signed "E. Hartnack // sucr. de G. Oberhaeuser" made in 1864. This mechanical solution for the stand became the benchmark for microscope makers between 1850 and 1870. (Timo Mappes)

therefore decided to devote himself once again to fundamentally on enhancing the optical quality of his instruments on the basis of scientific calculations. However, he could not do this alone. In order to break new ground in the field of microscope construction, he needed a highly skilled mathematician and engineer who was also willing to commit to the company over the long term – without any guarantee of commercial or scientific success. Could Zeiss really hope to find a suitable candidate in the small university city of Jena?

1 Axel Stelzner: *Carl Zeiss in der Jenaer Tagespresse (1847–1888)*, in: *Carl Zeiss und Ernst Abbe. Leben, Wirken und Bedeutung*, ed. Rüdiger Stolz, Joachim Wittig. Jena, 1993, p. 100.

2 Cf. Moritz von Rohr: *Zur Geschichte der Zeissischen Werkstätte bis zum Tode Ernst Abbes.* Jena, 1936, p. 19.

3 Cf. Horst Alexander Willam: *Carl Zeiss.* Munich, 1947, p. 47. – Von Rohr, ibid.

4 Allusion to Exodus 16:3: "Would to God we had died by the hand of the LORD in the land of Egypt, when we sat by the flesh pots, and when we did eat bread to the full; for ye have brought us forth into this wilderness, to kill this whole assembly with hunger."

5   Cited in: Paul Gerhard Esche: *Carl Zeiss. Leben und Werk*. Jena, 1977, p. 32.

6   Cf. Frank Wogowa: *Universität und Revolution. Jena und die 'hochschulpolitischen' Reformbestrebungen 1848*, in: *Die Revolution 1848/49 in Thüringen*, ed. Hans-Werner Hahn, Werner Greiling. Rudolstadt, 1998, p. 455.

7   Careful records were kept of the number of microscopes sold. The list of microscopes which is now kept in the ZEISS Archives (BACZ 7710 Fabrikationsliste einfache und zusammengesetzte Mikroskope, vol. 1, 1847–1881) provides the serial number of each instrument sold, the recipient, his place of origin, and the month in which the business was transacted.

8   Letter from Carl Zeiss to K. O. Beck dated 4 February 1855. Source: ZEISS Archives CZO-S 3.

9   Cf. Erich Zeiss: *Hof- und Universitätsmechanikus Dr. h. c. Carl Zeiss der Gründer der Optischen Werkstätte zu Jena*. [No place of publication] 1966, p. 29.

10 Cf. Letter from Carl Zeiss to the faculty of Philosophy at the Royal University of Greifswald, 18 August 1850. Source: ZEISS Archives CZO-S 581.

11 Letter from Eduard Baumstark to Carl Zeiss, 14 October 1850. Source: ZEISS Archives CZO-S 581.

12 Cf. also: Herbert Koch: *Unbekanntes über Leben und Wirken von Carl Zeiss*. [typewritten manuscript] Jena, 1957, p. 30. Source: ZEISS Archives BACZ 14605.

13 Cf. Willam, *Carl Zeiss*, p. 73.

14 Cf. Willam, *Carl Zeiss*, p. 9.

15 Letter from Carl Zeiss to K. O. Beck dated 4 February 1855. Source: ZEISS Archives CZO-S 3.

16 Cf. Willam, *Carl Zeiss*, p. 74.

17 Cf. on the importance of Fraunhofer: Dieter Gerlach: *Geschichte der Mikroskopie*. Frankfurt am Main, 2009, pp. 216–224.

18 Cf. ebd., p. 224.

19 Cf. obituary by Ludwig Kunze containing details of Barfuss's life, published by Moritz von Rohr in the journal *Deutsche optische Wochenschrift* 1927, no. 6, pp. 72–74.

20 Cf. the letter dated 20 March sent by Carl Zeiss to Joseph Ignatz Toepler, from an original publication held in the archives of Dresden University of Technology: Joachim Wittig: *Carl Zeiss und die Universität Jena*, in: *Carl Zeiss und Ernst Abbe. Leben, Wirken und Bedeutung*, ed. Rüdiger Stolz, Joachim Wittig. Jena, 1993, p. 26.

21 Willam (*Carl Zeiss*, p. 75) writes that Carl Zeiss and Friedrich Wilhelm Barfuss became acquainted in the winter of 1852–53; Paul Gerhard Esche (*Carl Zeiss. Leben und Werk*. Jena 1977, p. 35) states that their acquaintance began earlier.

22 Friedrich Wilhelm Barfuss: "Ueber die Construction zusammengesetzter Microskope," in: *Annalen der Physik* 68 (1846), pp. 88–91; here: p. 91.

23 In his 1888 obituary on Carl Zeiss, Abbe described the collaboration with Barfuss as a "complete and utter failure."

24 Matthias Jakob Schleiden: "Grundzüge der wissenschaftlichen Botanik (1842)," in: *Wissenschaftsphilosophische Schriften*, ed. Ulrich Charpa. Cologne, 1989, p. 130.

25 Carl Zeiss: "Ueber eine Erscheinung in Mikroskopen bei schiefer Beleuchtung der Objecte," in: *Annalen der Physik und Chemie* 13 (1858), pp. 654–656.

26 This information can be found exclusively in Dieter Gerlach: *Geschichte der Mikroskopie*. Frankfurt am Main, 2009, p. 346, which refers to the company's books. It states that the amount was reimbursed by Zeiss in December 1862.

27 Herbert Koch: *Unbekanntes über Leben und Wirken von Carl Zeiss*. [typewritten manuscript] Jena, 1957, p. 17. Source: ZEISS Archives BACZ 14605.

28 All the events that led up to Zeiss being named Deputy Master of Weights and Measures are detailed in: Koch, Unbekanntes, pp. 3–18.

29 Jena University Archives 471, sheet 4. P

30 Cited in: Koch, *Unbekanntes*, p. 28 and p. 30.

31 Cited in: ibid., p. 30.

32 Cf. lectures at the comprehensive university of Jena [digital resource, valid for the period 1866–1921: http://zs.thulb.uni-jena.de/receive/jportal_jpvolume_00215101] and view university course catalogs in the university archives from 1860 onwards.

33 ZEISS Archives CZO-S 727.

34 Printed in: Axel Stelzner: *Carl Zeiß in der Jenaer Tagespresse*, in: *Carl Zeiss und Ernst Abbe*, ed. Rüdiger Stolz, Joachim Wittig. Jena, 1993, p. 107.

35 Ibid., p. 109.

36 Cf. Hermann Koch: *Nachweis über die Einführung der Zeiss-Mikroskoptypen und der wichtigsten Nebenapparate von 1846–1900*. [typewritten manuscript] Jena, 1946, p. 4. Source: ZEISS Archives BACZ 1438.

37 Hermann Schacht: *Das Mikroskop und seine Anwendung, insbesondere für Pflanzen-Anatomie*. Berlin, 1862, pp. 24–25.

38 Cited by: Horst Alexander Willam: Carl Zeiss. 1816–1888. Munich, 1967, p. 35.

39 Letter from Carl Zeiss to schools inspector Dr. Eberhardt dated 9 August 1862. Source: ZEISS Archives CZO-S 532.

40 Transcript of Eidesnotul dated 6 October 1866. Source: ZEISS Archives BACZ 12260.

41 Cf. Gerlach, *Geschichte der Mikroskopie*, pp. 351–352.

Chapter 4

Putting Theory into
Practice:
The Major Shift to Scientific
Microscope Construction

« *Page 83: Carl Zeiss aged around 50, 1866.*

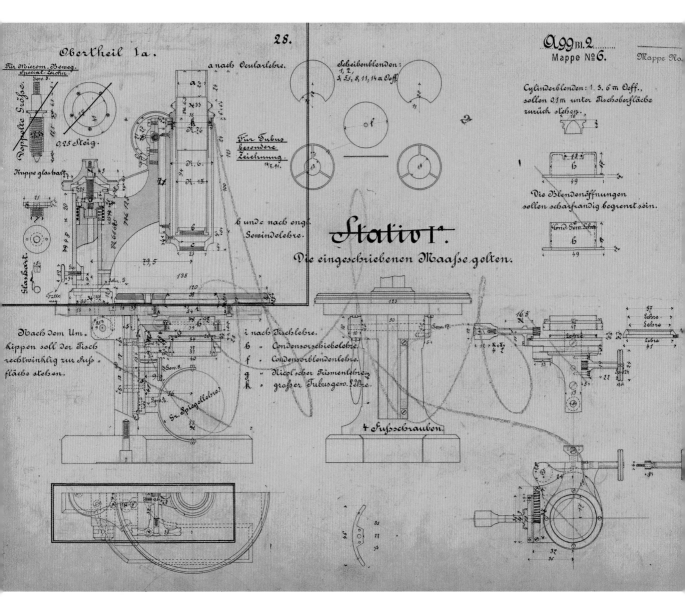

Sketch of the Ia microscope stand from the 1880s.

"I Have Never Seen More Beautiful Microscopic Images":
Optics Based on Calculations

"This is the idea that Carl Zeiss introduced to the world of microscope optics and realized despite all the obstacles that lay in his path: the idea of a highly rational setup for optical constructions pertaining to the microscope. It is the essence from which all internal progress and all external success triggered by his work in this respect were born. [...] Carl Zeiss could not have [...] achieved all of this on his own [...]. Due to the natural limitations of his own abilities, he was much more reliant on the contributions of others than Fraunhofer; his success therefore always hinged on the support of others. This does nothing to diminish his personal legacy. [...] It would not be accurate to say that his success was ultimately attributable to good fortune: he found [the] employees who were indispensable for his work because he <u>sought them out</u> – and tirelessly continued his quest in matters where others would have capitulated in light of the failures he sustained time and again. If indeed it is possible to speak of good fortune in the case of Zeiss, it is more the kind of good fortune that is associated with the adage 'every man is an architect of his own fortune'."[1]

When Ernst Abbe spoke these words to the employees at the company's 50th anniversary in 1896, Carl Zeiss had already been deceased for eight years. Abbe chose not to epitomize the deceased Zeiss as a scientific entrepreneur. This position was in fact already held by the much-admired Joseph Fraunhofer. After all, there were "limitations" as far as Zeiss's abilities were concerned. Abbe nevertheless conceded three things in relation to his erstwhile employer and subsequent friend: first, the idea of constructing

Ernst Abbe circa 1875.

microscopes on the basis of rational calculations; second, the perseverance needed to bring this idea to life; and third, the flair for recruiting extremely gifted employees and retaining them over the long term. Abbe was of course referring primarily to himself and – to a lesser degree – to the foreman August Löber. So Zeiss had "sought out" Abbe – but how did he actually find him?

1866 was the most successful year for Carl Zeiss since he established his company. On 28 May he celebrated the production of the 1,000th microscope with his 11-man workforce. In a horse and cart, they all made their way to the small town of Tautenburg, which lies some twelve kilometers outside Jena; time for celebration, however, was limited. 192 microscopes were produced by the company in 1866 – 81 more than in the previous year, even though the workforce had only grown by one. Under Löber's supervision, the Jena Optical Works attained a higher standard of production. As far as the quality of workmanship was concerned, Zeiss had risen to the upper echelons among European manufacturers. He utilized the momentum afforded by a strong fiscal year to invest in the future of his company. The key move was his decision to hire, at long last, a scientist who could help him find solutions to the all-too-familiar prob-

Zeiss men's choir: August Löber with his employees at the Zeiss workshop in 1869. From left to right (standing): Carl Eisenhardt, Joseph Rudolf, August Löber, unknown person (probably the director), Fritz Töpfer, Carl Schäfer; From left to right (seated): Wilhelm Böber, Fritz Müller, Carl Müller, Heinrich Pape.

lems associated with optical production. He opted for 26-year-old Ernst Abbe (1840–1905), who had earned a good reputation at the University of Jena as a private lecturer for mathematics and physics. Abbe, whose father was a foreman at the Eisenach spinning mill run by the Eichel-Streibers, owed his scientific education and career to his outstanding achievements and to the generosity of several patrons. Given the

meager salary of a private lecturer in Jena, he was no stranger to financial hardship. It is likely that Zeiss and Abbe met in 1863 when the former produced a measuring instrument for a physics lecture given by Abbe. As Abbe had already studied in Jena from 1857 to 1859, they may have already met during this period, even if only briefly.

Zeiss and Abbe join forces

Historians claim that the partnership was officially sealed on 3 July 1866. Just as with the exact date on which the company ZEISS was established, documentary evidence from the time is also lacking in this regard. The date was reconstructed based on Abbe having celebrated his 25th anniversary with the company on 3 July 1891. There are no written agreements dating back to 1866, meaning we can but make assumptions as to the conditions and objectives that ultimately persuaded Abbe to work with Zeiss. What we do know, however, is that Abbe was not the obvious choice to fill a position as a scientific advisor. The young physicist and mathematician had previously concerned himself only in the slightest respect with matters pertaining to optics. The focus of Abbe's doctorate was on the first law of thermodynamics; his postdoctoral thesis concerned the least squares method with which Carl Friedrich Gauss (1777–1855) is credited, i.e. with a general mathematical method for error correction and the optimization of measuring series. What is remarkable is that Abbe and Zeiss, who made a significant contribution to the advancement of the optical industry in the 19th century, were newcomers to the field. If we are to believe Siegfried Czapski (1861–1907), who was Abbe's closest colleague in optical development in Jena from 1886 onwards, Abbe found it relatively easy to break new ground in microscope production because he had not received a conventional education on the subject:

"E. Abbe […] possessed […] no more than the basic tools of a mathematician and physicist, and even

in this respect was […] severely lacking, one might say thoroughly ill-informed. Perhaps it helped that, upon commencement of his work, he had no inkling of the challenges that lay ahead; perhaps his […] insufficient education on the subject laid good foundations in that it enabled him to adopt an unbiased approach to the problem at hand."[2]

Siegfried Czapski.

The 'problem,' as Czapski describes it, began for Abbe even before he attempted to develop a new theory about the microscope. The overarching issue was how to turn the Works in Jena into an efficient production facility with little wastage and a consistently high level of quality. Abbe took a critical look at the established practices of the Zeiss workshop and began to suggest a vast array of changes – at first very much to the annoyance of the foreman August Löber and other long-standing workers. However, Carl Zeiss was clever enough not to assert his authority as head of the company over the new recruit, 23 years his junior, and not stand in the way of meaningful change. Abbe described his behavior toward Zeiss at that time in the following way:

"[Carl Zeiss] was always agreeable, empathetic and friendly to all those who worked with him; but he

Bird's-eye view of the World's Fair in Paris, 1867.

also had very high expectations of them all because he was accustomed to expecting no less of himself. But it was not by admonishing and reprimanding his workers that he enforced his will; endowed with a natural wit, he preferred to manage his workforce with a little mockery and irony, softening the blow with his affable nature. He treated me [...] in the same way, as a father figure and friend, when I, as a very young man and wholly inexperienced, entered his sphere of influence.

But what raised him in people's estimations alongside his personality were his pronounced sense of duty and highly developed sense of justice. There are a number of examples to support my statements, but I shall limit myself to instances that affected me per-

sonally: the liberal and selfless way in which he constantly made efforts to retain me as an employee. It would never have even crossed his mind to use to his own advantage my reliance on him, having neither a fortune of my own nor other such succor in life."[3]

Zeiss was not merely content with looking for ways of expanding his company in his immediate environs; this is demonstrated by his visit to the World's Fair held in Paris in 1867.[4] While there are no records pertaining to possible business contacts he might have made, it is well known that numerous major competitors were represented at the event.[5] It therefore seems reasonable to assume that Zeiss gained an impression of the latest developments in technology across the world.

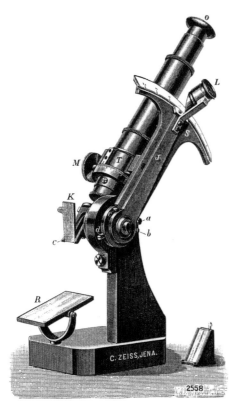

Abbe-type refractometer with non-heatable prisms, second model, 1904.

Not leaving anything to chance

Meanwhile, Abbe had devoted his attention to developing an array of measuring instruments to precisely determine the optical properties of lenses – an important prerequisite for efficient production. He began in 1867 by presenting a focometer for measuring focal lengths; by 1870 he had also unveiled a spherometer to determine the outer radii of lenses, a thickness gauge, a refractometer to determine the refractive index and an apertometer to measure what was known as the numerical aperture, which was crucial for measuring the resolving power of optical instruments.[6] The long list of newly developed measuring instruments raises the question of how Carl Zeiss and his competitors had previously been able to produce

precise lenses without all of these possibilities. For traditional optical production, however, it was this lack of precision that provided the scope for measuring methods to be devised on the basis of intuition and luck. Skilled craftsmen like Löber, Hartnack (Paris) and Plössl (Vienna) used these methods to construct exceptional instruments.

In this sense, Zeiss's efforts to achieve greater accuracy and a consistently high level of quality posed a dilemma that was tantamount to a competitive drawback: while others took an open-minded approach to trying out different lens combinations and were successful time and again, Zeiss imposed restrictions on his own work by attempting to rationalize production.[7] By recruiting Abbe, Zeiss was able to advance this process in spite of both internal and external resistance. What is more, the scientist provided him

The Abbe apertometer, used as of 1870 to measure large apertures (opening angles).

with a kind of dynamism that surprised even himself. By 1857 Zeiss had taken the first step toward a 'division of labor' by separating optical and mechanical production. But Abbe was not content to stop there and pushed for a more definitive division of individual stages of production. The latter part of the 1860s can therefore be described as a transitional phase for the Zeiss workshop in which the company was subject to temporary setbacks in order for fundamental processes to become established. This may be why the company did not instantly flourish in spite of its numerous innovations. Zeiss had also not yet managed to navigate the economic peaks and troughs of the past few decades and bring stability to his income. In 1867 microscope sales fell by more than half compared to the previous year. In 1868 the workshop received considerably more orders, only to experience another deep slump in 1869.

More efficiency in production

At long last, the collaboration with Abbe, which had already generated substantial costs, yielded its first notable success: from 1869 onwards, Zeiss drastically lowered the price of most microscope lenses, offering reductions of up to 25 percent. This was made possible by modernizing production; new measuring instruments were introduced and processes revised. Largely unbeknown to customers, a process of innovation had also been launched in the Zeiss workshop that would change microscope technology from the ground up. Abbe had built a new kind of illumination apparatus that was mounted below the slide and

Catalog no. 19 from 1872.

resembled an inverted microscope lens. This 'condenser,' which became available in 1869, was actually built to study microscope aberrations. As this precise focusing of the illumination also held immense value for scientific work, the illumination apparatus was included in the official price list from 1872 onwards. It was probably unclear to most people at the time why the 'condenser' with its large aperture was superior to the concave mirror previously used on virtually every microscope. Abbe had not yet published the theory

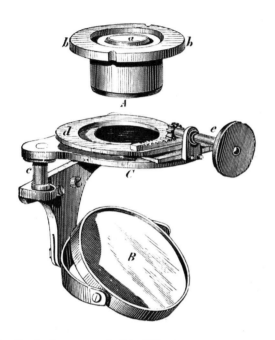

First illumination apparatus by Abbe, 1872.

Robert Koch in Kimberley in 1896 working with a Zeiss microscope, photographed by Frank H. Hancox.

behind the apparatus. Even so, virtually everyone who picked up one of these "little instruments"[8] was amazed. In 1878, it became clear to what extent Zeiss was ahead of his time by virtue of his collaboration with Abbe when the physician and subsequent Nobel laureate Robert Koch (1843–1910) stressed in his experiments on the etiology of wound infection that he had only been able to make his revolutionary breakthroughs using the apparatus built in Jena:

"An illumination cone had to be used that had an aperture large enough to ensure that diffraction phenomena disappear without a trace. After having tried a range of different lenses and condensers of this type without encountering a piece of apparatus that almost completely removed these phenomena, I finally discovered an instrument fully suited to my purpose in the illumination apparatus built by Carl Zeiss in Jena and designed by Abbe."[9]

1869, a year in which Zeiss built no more than 78 microscopes, appears to be a year of crisis as much as it does a turning point in the company's history. In the same year, Zeiss gave his scientifically minded partner Abbe a research task he had more or less also assigned to Friedrich Wilhelm Barfuss more than 15 years before: to construct a microscope lens based on calculations. Abbe was tasked with developing a system that was at least as good as the water immersion objective lenses offered by Edmund Hartnack which, at the time, were hailed as the best lenses in the world.

Unforeseen difficulties

Abbe had soon come up with a design that corrected many of the known aberrations. When it came to testing his theory, however, the benefits that looked good on paper were nowhere to be seen. The lens, which was produced to a precise mathematical formula and to the highest quality standards in the Zeiss workshop, demonstrated a much lower resolving power than its 'unscientific' predecessor produced through trial and error. Abbe had clearly underestimated his new task and began to suspect that microscopes worked on principles which, to some extent, were as yet undiscovered. Abbe had spent the better part of a year working flat out with nothing to show for his labors, and still Zeiss did not call their partnership into question. With funding still coming his way, Abbe was able to put his hypotheses to the test time and again by carrying out measurements and optical experiments and by comparing the Zeiss microscopes with all others on the market. Such scientific trial and error ultimately enabled him to bridge the gap between the lack of experience at the Zeiss workshop and the long-established competitors, and subsequently surpass them.

A biography on Carl Zeiss is not the place to trace the development of Abbe's optical theories with all of

Calculations by Abbe on the back of a letter.

their ramifications. Nevertheless, we should mention the major innovations that characterized the period from 1869 to 1872:

- Abbe recognized that a large aperture enhanced the resolution achieved with microscope lenses even though his theory in fact posited the opposite.
- This contradiction caused Abbe to advance from established geometrical optics to wave optics. In part, therefore, he abandoned the idea of rays

of light as geometric lines in favor of the modern characterization of light as waves; this caused him to break away from the Newtonian theory that light is made up of particles.

- Abbe now understood the effect of diffraction and could explain why lens resolution was determined by the aperture (i.e. the size of the opening through which light falls).
- The Zeiss workshop was now in a position to produce medium-power lenses based on calculations. But still the Zeiss workshop did not come close to the industry's finest optics produced in the traditional manner.
- The missing piece of the puzzle was the "sine condition" that Abbe probably came up with in the fall of 1870 and which was later to bear his name. At long last a mathematical formula for the image quality of microscope lenses had been found.

Production is restructured

On 12 September 1871 Abbe presented his design plan for a powerful water immersion objective lens. The breakthrough Zeiss had been eagerly awaiting for two decades was finally achieved. The task now was to base the Jena workshop's entire product range on the new theory. In price list no. 19, published in 1872, Zeiss announced the project to be a success: "The microscope systems presented here were all built recently on the basis of the theoretical calculations made by Professor Abbe in Jena."[10] For the first time since 1857, Zeiss unveiled new objective lens systems, including some with larger apertures

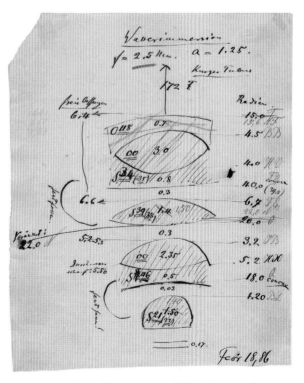

Immersion lens K 1881 (Timo Mappes) and Abbe's calculations for a system of this kind.

and three different water immersion systems.[11]
The botanist Leopold Dippel, who reviewed Zeiss's
portfolio, was among the first to use these systems.
In an essay entitled *Die neuen Objektivsysteme von
Carl Zeiss*, which was published in November 1873 in
the magazine Flora, the eminent microscopist of the
Jena-based workshop penned a glowing review:

*"Despite being able to surmise from earlier admira-
ble achievements made by Zeiss that the announced
lens systems would certainly deliver outstanding
performance, during examinations I made of a large
number of systems bearing the same number which
came fresh from the workshop, I was nevertheless
surprised to discover a device that met my every
need in a manner I had hitherto never dreamed
possible and in a way that can only be accomplished
through strict adherence to a theoretical basis.
[…]
The performance when examining different organic
specimens […] was exceptional, and I can assure you
that I have never seen more beautiful microscopic
images of the finest and most difficult structures
than those created by the Zeiss lenses, be it with
central or oblique illumination."[12]*

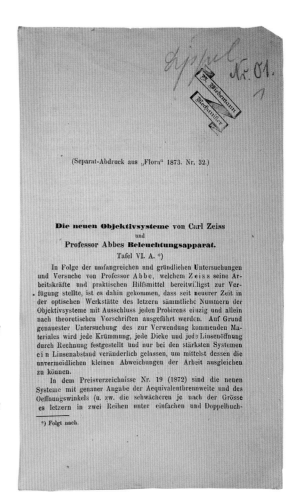

Essay by Dr. Leopold Dippel from the magazine "Flora", 1873.

The consistency of the optical properties noted by
Dippel was indeed a great advantage afforded by the
new microscopes based on calculations – particularly if several laboratories were working simultaneously on one issue. In this respect, Zeiss could offer
something truly unique that was of particular interest
for more advanced research undertakings. Observations made using a microscope became much easier

to reproduce and understand – as long as all those
involved used products from Zeiss. No longer were
there any great discrepancies between microscopes
in a single product range, something that was practi-
cally unavoidable with optics produced on the basis
of trial and error.

A new system is born

While Zeiss supplied the scientists of the time with powerful devices, his scientific companion was rewriting the theoretical rules for the optical industry as a whole. In the essay *Beiträge über die Theorie des Mikroskops und der mikroskopischen Wahrnehmung*, Abbe, now an associate professor, summarized the findings he had made in recent years. He stated in no uncertain terms that the progress made in the field of optics had only been possible through a combination of theoretical and practical aspects:

"By virtue of the willingness of Mr. Zeiss […] which he demonstrated to me by placing excellent auxiliary equipment and his diligent workforce at my disposal over a period spanning several years, and as a result of the alacrity of such a skilled foreman and his competent apprentices, who accepted the incoming work, our efforts have finally borne fruit. […] The relevant constructions are […] supported by calculations down to the very last detail – every curvature, every thickness, every lens aperture – so much so that all trial and error has been abandoned. […] This shows that a sufficiently detailed theory, combined with efficient technology that draws on all auxiliary equipment physics has to offer for practical optics, can also successfully replace empirical procedures in microscope construction."[13]

Abbe and Zeiss had managed to revolutionize a system: the Jena-based Optical Works with their shop was poised to develop into a completely new kind of scientific enterprise.

1   Ernst Abbe: "Gedachtnisrede zur Feier des 50-jährigen Bestehens der Optischen Werkstätte (1896)", in: *Gesammelte Abhandlungen vol. III.* Hildesheim, 1989, pp. 65–73.

2   Cited by: Dieter Gerlach: *Geschichte der Mikroskopie.* Frankfurt am Main, 2009, p. 406.

3   Abbe, commemorative speech, p. 77.

4   This information is based on a statement made by Zeiss's daughter Hedwig to Moritz von Rohr. Cf. Moritz von Rohr: *Zur Geschichte der Zeissischen Werkstätte bis zum Tode Ernst Abbes.* Jena, 1936, p. 19.

5   Cf. Horst Alexander Willam: *Carl Zeiss.* Munich, 1967, p. 105.

6   Cf. Gerlach, *Geschichte der Miroskopie*, p. 406.

7   Abbe described this dilemma in retrospect. Cf. Abbe, commemorative speech, Appendix 2, p. 98.

8   According to Leopold Dippel in a letter to Ernst Abbe; cited by: Gerlach, *Geschichte der Mikroskopie*, p. 400.

9   Robert Koch: *Untersuchungen über die Aetiologie der Wundkrankheiten.* Leipzig, 1878, p. 34.

10 Cited by: Gerlach, *Geschichte der Mikroskopie*, p. 418.

11 Cf. Herbert Koch: *Nachweis über die Einführung der Zeiss-Mikroskoptypen und der wichtigsten Nebenapparate von 1846–1900.* [typewritten manuscript, Jena] 1946, p. 6.

12 Leopold Dippel: *Die neuen Objektivsysteme von Carl Zeiss und Professor Abbes Beleuchtungsapparat,* published in *Flora or Botanische Zeitung* 32 (1873), pp. 497–498.

13 Ernst Abbe: *Beiträge über die Theorie des Mikroskops und der mikroskopischen Wahrnehmung in: Abhandlungen über die Theorie des Mikroskops. (Gesammelte Abhandlungen II)* Hildesheim, 1989, pp. 46–47.

Turning Art into Science:
The Past, the Present and the Future of Microscopy

Dr. Timo Mappes (Head of Research & Development at ZEISS Vision Care and collector of historical microscopes, pictured left) talks to the physicist and Nobel laureate Dr. Eric Betzig.

Mappes: Looking back on the person Carl Zeiss from your perspective as a scientist today, which aspects grab your attention?
Betzig: Honestly, I am not an avid student of the past and it has been a few years since I read a historical book from start to finish. However, if I had a time machine, I would like to visit Carl Zeiss during the period when building microscopes was still an art. I imagine that he was frustrated because his goal was to achieve volume production of high-resolution

microscopes with a consistent quality. Then Zeiss met Abbe – one of my personal heroes, by the way – and together, they developed a mathematical theory that turned the art of making microscopes into a science. Still, they were unable to really get what they wanted as the lens material had too many flaws. They needed to get Schott involved in order to make a commercial breakthrough. This is how it happened: one of the great success stories of combining pure and applied science.

Mappes: So you see this story as an early example of interdisciplinary team-play …
Betzig: Exactly, there is a siloization in science that is often problematic today, but was less typical in

the 19th century. Carl Zeiss and his colleagues were interdisciplinary without even using the term. Instead of considering themselves physicists or engineers or biologists, they were more application-driven. Taking input from a plethora of sources, they learned what they had to learn. I think that is something we should aspire to, and we do that with our projects at the Janelia Research Campus.

Looking back at Carl Zeiss, you find that he was not an expert in every field important for his goal. Yet he was ready to work outside his comfort zone, and he knew how to identify the best people and bring them together. And there is another thing: He was willing to enter new fields of research, even though that involved financial and personal risks.

Mappes: Speaking of risks, there is always a certain probability that a successful development will turn out to be a commercial failure …
Betzig: I guess that is the nature of business. You have to balance potential gains and risks, like in a poker game. If you end up being too conservative, you will never be a good poker player. If, however, you manage to fail sometimes, but also to, when you win, cover the cost of your failures and still gain something, you start to advance. I do not think that the decision about whether or not to move from the lab to the factory floor can ever be fully rational, but I am sure that commercialization of technological advances is essential.

I have worked in business at several stages of my career, so I know how "R" and "D" interact within the framework of research and development. Imagine a pyramid: research is on top, development is at the base. You cannot get anywhere without the base of the pyramid. If you do not make that effort, your research will be far too limited to have any lasting impact. As a good scientist, you know that in the end, you will not be judged by the metric of papers, but by whether your findings make a difference. If your area of work is technology, that requires commercialization. Nowadays, however, we live in a time of rapid change. In the 21st century, microscopy is in a state where many technologies are ephemeral. This also becomes a commercial risk, as you can be at the halfway point in a product development cycle – and then find out that the product is obsolete. It is a stressful but exciting time.

Mappes: Going back to Carl Zeiss and his time: Have you ever had the chance to use one of those antique instruments built by his workshop in the 1846 – 1888 period?
Betzig: Yes, I have. I was given one of the early compound microscopes and I looked through it. It is truly impressive to see how far you could go without any modern-day equipment. This reminds me of another pioneer of microscopy, the Dutchman Antoni van Leeuwenhoek who, towards the end of the 17th century, had already discovered bacteria. Leuwenhoek was maybe 200 years ahead of his time; but unfortunately, much of his progress was lost because he refused to disclose his method of lens making.

Mappes: Actually, there seem to be many ways in which progress in microscopy stalled. What

strikes me is how often previous developments fell into a kind of state of hibernation until, several decades later, they were eventually revived …

Betzig: This is certainly true for plane illumination microscopy, going back to the ultramicroscope that Zsigmondy and Siedentopf built in 1912. I also view adaptive optics in a similar way: This technology has been used by astronomers for a long time, but only very recently have microscopists started to understand its potential. We have to adapt it to the specific conditions of biological samples though. I daresay that, in less than 15 years from now, people will probably not buy a set of individual objectives to deal with different types of aberrations and samples, but that we will rather have one objective that is able to adapt to immersion, sample type, etc. This seems to be a phenomenon that you often see in the industry: a trend towards complexification followed by an opposing drift towards simplification.

Mappes: When fluorescence microscopy was discovered by August Köhler at Zeiss around the turn of the 20th century, this was a discovery mostly by chance. Does that sound familiar to you?

Betzig: Absolutely, when you are researching into a new field, there is always an element of trial and error involved. Often enough, just when I got the impression that I had hit a wall, all of a sudden, I would get amazingly lucky. But then again, failure does not mean that you are standing still. As Thomas Edison said of inventing the light bulb: he had to find the 999 ways that did not work first.

Mappes: Besides being a scientist, you have also worked as head of R&D at your father's Ann Arbour Machine Company. What do you think about the relationship between science and industry?

Betzig: In my opinion, one of the biggest mistakes is that research into microscopes often starts in isolation. This is the inverse of what should be: Research should be driven by the customer. That is when you think about how to find a solution for what your customers need by building a specific type of microscope. As a company, you have to have a clear idea of the market if there is going to be any incentive to make large investments. You need those large investments because your goal should be to manufacture a turn-key product that has an impact on the end user. In other words, scientific instruments, i.e. microscopes, should be like a simple box that the users can forget about so that they can concentrate on science. A microscope, just like any other product, should offer high performance while remaining cost-effective.

Mappes: When deciding which path to take in R&D, most manufacturers will usually stick to the technologies that they understand best. However, if what you say about the customer as a driving force is to be taken seriously, we should also be open for more radical new ideas …

Betzig: My point is that you have to look at the whole chain, just like Carl Zeiss did when leaving behind the traditional method of making microscopes in favor of a more scientific approach. Looking further ahead may set you back in the beginning, but will help you to success in the long run. When I was in graduate school, I heard a talk by Edwin Herbert Land, the man behind Polaroid. He likened the goal of creating a marketable product to a ladder with missing rungs. Many people only look at the ladders that have all the rungs at the bottom, even if there are twenty missing rungs further on. The trick is to look at the whole ladder – and to make sure that there are not too many rungs missing in a row so that you can work your way up.

Mappes: When you think about going up the ladder, what would you say lies at the top? What is going to be the next Nobel Prize in the field of microscopy?

Betzig: I would place a bet on cryo-electron microscopy. This method allows structural biologists to conduct research on an essentially atomic level, which is quite simply revolutionary. Then there are some other things which might be considered the 'holy grails' of microscopy today: live imaging at high resolution, label-free methods capable of replacing fluorescence, or the ability to study the interactions of, say, a few hundreds of proteins at once, instead of looking at just two or three at a time.

In a way, the field of super-resolution microscopy is similar to the early days of the automobile industry: We are going to see a lot of refinement in the years to come – and a lot of shake-out.

....

Dr. Robert Eric Betzig (*1960 in Ann Arbor) studied physics at the California Institute of Technology and at Cornell University, where he graduated with a PhD. He has worked at different research institutes and for a family company. Since December 2005 he has been heading up a research group at the Janelia Farm Research Campus of the Howard Hughes Medical Institute. In 2014 he was awarded the Nobel Prize in Chemistry for the development of super-resolved fluorescence microscopy.

An Investment in the Future:
The Rise from Optical Works to Enterprise (1873 – 1880)

Zeiss and Abbe put microscope optics on a new basis – and it was nothing short of a revolution. What is surprising is the steady course of the company's development in the 1870s. There were no drastic rises in sales figures or personnel. In fact, the two men were able to navigate the financial peaks and troughs of recent years and create a business model characterized by constant yet moderate growth. Zeiss also used this newly acquired financial stability to offer his employees a better kind of social security. This led him to support the establishment of a company health insurance plan on 1 January 1875, which offered its members such benefits as free treatment at a specific approved doctor and the reimbursement of medication fees, as well as financial support for a period of up to twelve weeks in the event of an inability to work.[1] As such, the transformation of the Jena

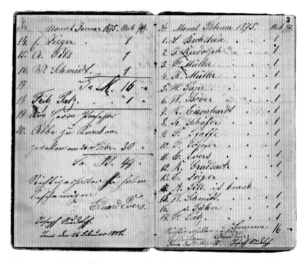

First page of the health insurer cash book, with the incoming monies for 1875.

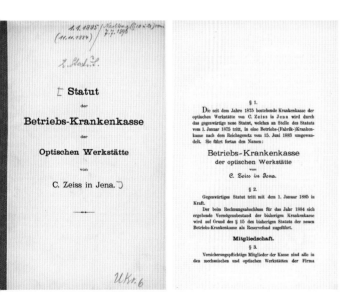

Company health insurer statute, 1885.

Optical Works into one of the most socially-minded companies of its time was prompted by Zeiss himself.

A further notable aspect is the clientèle of the Jena Optical Works, which became increasingly international. While it is true that Zeiss also supplied customers abroad with his first microscopes, for example in Prague, Cracow, Copenhagen and St. Petersburg, the majority of them between 1847 and 1869 were local, based in the cities of Jena, Weimar, Berlin and Leipzig. Zeiss continued to supply this market, but was also able to expand his client base to include England, France, the Netherlands, the USA and even Japan. For customers all over the world, the purchase of a Zeiss microscope was much more a conscious decision than it was for a researcher in Jena, especially given the

Page from balance sheet book no. 1 with customer name, including Prof. Dippel (no. 22), 1875.

Prophecy of doom?

Be that as it may, many of his contemporaries were skeptical about the transition from trial and error to optics based on calculations. This was particularly true of Zeiss's competitors. In London, Paris and Vienna, people were very reluctant to admit that a small workshop in a provincial town in eastern Germany had managed to outdo all other microscope builders on the continent. Ernst Abbe described this defiance among a portion of the established competitors as follows:

"Until not very long ago, when purchasing a micro-scope many people were attaching greater impor-tance to microscope-makers who upheld the old em-pirical method and justifying their decision by stating that these microscopes were not constructed in the same way as in Jena. Only in the last 10 years or so [since 1886] has the opposite assertion – construc-tion will take place just like it does in Jena – become a widespread means of bolstering claims of higher quality [...]." [2]

Initially, independent observers did not want to believe that the innovations launched by Zeiss did in fact herald a new era in the construction of precision optics. The physician Friedrich Merkel (1845–1919), for instance, warned against overestimating the inno-vations that came from Jena and stressed that "while the latest microscopes produced by Zeiss on the basis of calculations are very good, they are not superior to other microscopes not built in line with Abbe's

high-quality producers of microscopes in England and France. We can see that Zeiss succeeded in making a name for himself among the scientific elite across the globe.

Abbe becomes a shareholder

Such ventures into the realm of scholarly discourse were rather few and far between for Zeiss. As a company owner, he focused primarily on his own business – an area in which the 1870s were destined to bring significant change. If historical records are to be believed, it all started when the entire Abbe family contracted typhoid in 1874. The cause of the disease was still unknown at the time and it was not until 1880 that it was discovered by the pathologist Karl Joseph Eberth. The lack of a viable therapy meant that Abbe had to fight the terrible infectious disease that causes people to run a fever, thus resulting in a protracted inability to work and high medical fees. His savings were depleted in no time at all. In April 1875 Abbe therefore wrote a letter to Zeiss in which he asked for a share of the profits generated by optical production. Zeiss responded on 19 May 1875 with an offer that far exceeded Abbe's expectations. Below is a lengthy excerpt that also illustrates Zeiss's vision for the future of the company he founded.

"Most esteemed Professor Abbe,

You chose the written form to notify me of your wish to change the nature of our partnership; I agree that this is a fitting way to converse on the most salient points of our future collaboration.
[…]
On the basis of your suggestion that, assuming an appropriately opportune course of business, […]
you would be entitled to approximately ⅓ of the net profits of the Optical Works, I am certain that you will

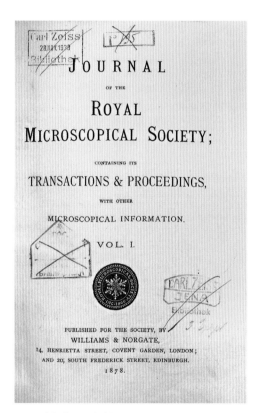

JOURNAL

OF THE

ROYAL
MICROSCOPICAL SOCIETY;

CONTAINING ITS

TRANSACTIONS & PROCEEDINGS,

WITH OTHER

MICROSCOPICAL INFORMATION.

VOL. I.

PUBLISHED FOR THE SOCIETY, BY
WILLIAMS & NORGATE,
14, HENRIETTA STREET, COVENT GARDEN, LONDON;
AND 20, SOUTH FREDERICK STREET, EDINBURGH.
1878.

Cover of the "Journal of the Royal Microscopical Society," 1. 1878.

calculations – in actual fact, in several significant respects"[3]. In contrast, the new developments triggered by Zeiss were received more favorably by the Royal Microscopical Society in England. The mathematician John Ware Stephenson (1819–1901), with whom Abbe and Zeiss both corresponded, fervently supported this turning point for microscope optics based on calculations. The friendly relations with England were also underscored in 1877 when Zeiss presented the apertometer built in his workshop in the magazine published by the Royal Microscopical Society.[4]

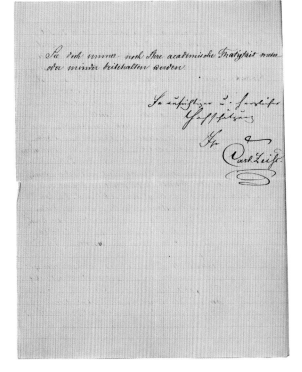

Letter from Carl Zeiss to Ernst Abbe dated 19 May 1875, (Pages 1, 2, 5 and 6).

view my offer of one third of said net profits of the entire business as very reasonable […].

[…]

I attribute my proposal, firstly, to my wish that your outstanding talent not only in rebus mathematicis but also in the construction of apparatus be assimilated in and bound with the business. My proposal is also based on careful reflection that it is presently well nigh impossible to determine the net profits of the Optical Works.

[…]

As regards this suggestion that names you my partner and entitles you to a third of the net profits, I am all the more convinced of your approval in that, during our previous negotiations, you deemed it appropriate to calculate the share of profits stipulated at the time from the systems that were veritably <u>sold</u> and not only produced.

[…]

Hence, it is imperative that we introduce a wholly different and purely financial management function and, to this end, must employ a learned and capable young financial expert, whom I intend to replace with my son Roderich in 2 or 3 years. Roderich has agreed to my suggestion, of which I informed you previously, and will join a larger factory business, perhaps in Rathenow, to acquire a thorough schooling in financial matters.

[…]

It would therefore stand to reason for me to conclude this agreement on behalf of the company for a set number of years so that the business may be continued in the event of my demise, at the very least for the period set forth.

As concerns the explanation you requested about the proprietorship of your mathematical calculations, you may choose to forgo the proposal of your own accord. Alternatively, I would ask, my dear Professor, that you consider what costs have been incurred in the realization of said calculations. First of all, your scientific acumen and your four years of service, second your dedication to my optical and in part mechanical workshops over four years, and a number of minor payments made to you in cash which were small compared to the wages I pay my employees who are working on our experiments, not to mention some other procurements. Please forgive me, esteemed Professor, as I cannot yield to you on this point; I believe that the amounts in question are a product of the work we have done together.

[…]

I would ask that you take some time to consider my proposal at your leisure. At present, I do not think the world needs to know that you are my associate; I will continue to devote all of my time, from morning to night, to the business and you may more or less continue to observe your academic commitments." [5]

Zeiss did not simply make Abbe an offer he could not refuse – he also amended the conditions of their partnership in a way that better safeguarded the strategic interests of the company. Contrary to Abbe's declared wish, i.e. to only have a share in the profits linked to his own developments, Zeiss offered him a holding in the entire company – including the mechanical workshop and retail business. On first glance, this meant more income, but it also implied a participation in potential losses resulting from business activities

Double-page spread from cash book no. 1 (1875) with revenue and expenditures.

that were not directly related to optical production. In a sense, Zeiss gave Abbe an incentive to combine his own interests with those of the entire company. In much the same way, Zeiss objected to the "proprietorship of Abbe's mathematical calculations", in other words the rights conferred to Abbe in the context of the optics developed by Zeiss. Even though Abbe's letter was not preserved, Zeiss's words suggest that the initial request was for these rights to be transferred to Abbe. Zeiss, however, insisted that the developments were the product of joint efforts and, as such, were owned by the company.

Roderich Zeiss – an unwilling successor

58-year-old Zeiss was well aware that he would retire much earlier than Abbe, who was 23 years younger. As the latter had been focusing on product development and optimization in production and was still pursuing his scientific endeavors, Zeiss also deemed it necessary to recruit a director who was well-versed in financial matters. Zeiss planned for this position to be held by his eldest son Roderich. His plan, however, was favored more by him than by his son. Roderich had in fact wanted to become a surgeon, but after being severely injured in the Franco-Prussian War, he suffered paralysis in his right hand. His dreams of becoming a surgeon were dashed. But this did not prevent him from graduating with a PhD in medicine from the University of Jena in 1875. Roderich had

certainly not set his sights on a career in business. As his father had preordained, in the summer of 1875 he therefore began an apprenticeship at the company Emil Busch Aktiengesellschaft in Rathenow, one of the major German producers of eyeglasses and similar instruments. Zeiss Senior was a long-time acquaintance of Emil Busch (1820–1888), who had taken over the company in 1845 from the microscope constructor Eduard Duncker (1767–1843), among other things as a supplier to his retail business. In 1859 he had described Busch as "exceptionally punctual, respectable and conscientious in every way."[6] There was no doubt that Roderich would be in good hands at such a model company. Contrary to his father's comments in the letter to Abbe, Roderich returned to Jena after just one year to involuntarily pursue a career as a businessman and not as a

Roderich Zeiss circa 1870.

The building of the Optische Industrie-Anstalt (optical industrial enterprise) in Rathenow, where Roderich Zeiss completed his traineeship, circa 1880.

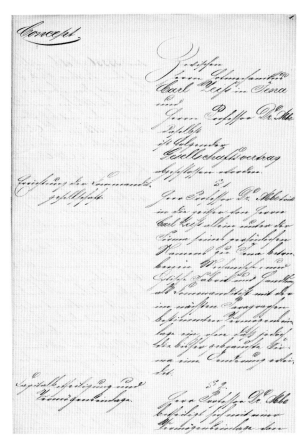

Draft of the partnership agreement from 1876 between Ernst Abbe and Carl and Roderich Zeiss.

bled Zeiss to fulfill his ambition of ensuring Abbe was completely committed to the company. The details were finalized in a partnership agreement dated 15 May 1875, which was not signed until the summer of 1876 at the earliest.[7] As per the agreement, Abbe was not, as initially intended, entitled to a third of the business, but became an equal partner. Nevertheless, net profits in the first five years were to be divided between Zeiss and Abbe at a ratio of 3:2. There were no plans for the weighting to be amended in the Zeiss family's favor after Roderich Zeiss joined the company: while Roderich was to assume equal responsibility for the management of the company, he would not be entitled to his father's shares until the latter bowed out.

It was prohibited for the partners to work for a competitor or to divulge any business secrets. In addition, Abbe was contractually bound to compile his records in comprehensible form and to bequeath them to the company in the event of his death. He expressly refrained from accepting a possible appointment as a *professor ordinarius* which, unlike the post of 'professor extraordinarius,' would have meant considerably more teaching duties. Consequently, Zeiss could count on Abbe prioritizing their joint enterprise over his scientific activities.

From patriarch to partner

The agreement also contained a number of provisions designed to ensure that the company would continue to exist in the event of a temporary rift

physician like he intended, and joined his father's enterprise in the fall of 1876 at age 26. He officially became a co-owner on 1 October 1879.

Returning to the negotiations between Zeiss and Abbe: both agreed on a course of action that spelled significant financial benefits for Abbe and a strong position in the company's management. This ena-

The employees celebrate the 4,000ᵗʰ microscope in 1879.

between the parties or if one partner chose to leave. Zeiss was taking efforts to safeguard his legacy as an entrepreneur; this was also in the interests of his workers, of which he now had more than 20. For the first time, double-entry bookkeeping was established as an obligation, as were regular stocktaking and the recruitment of a bookkeeper. Furthermore, private and business assets were separated; Zeiss achieved this by 'selling' all machines and all stock to the company and 'letting' the workshop rooms – that

were previously part of his private residence – to the company in the standard way.

This emancipatory process benefited the company, which was now classed as a separate legal entity; this is demonstrated by the dynamic development noted in the years that followed. Putting production on the basis of precisely calculated optics was merely the beginning of an ongoing process of innovation. Not only was the image quality enhanced, but the microscopes

Immersion lens from 1880 (Timo Mappes).

also became easier to use. In 1877, for example, Zeiss unveiled a new range of stands with a tiltable upper part that could also be equipped with pinion drive adjustment and a lens turret to enable users to quickly switch between different lenses. For the first time, day-to-day work with a light microscope was possible in a way that approximates today's standards. The first homogeneous oil immersion objective lens became a veritable sales driver and was developed on the suggestion of the aforementioned John Ware Stephenson; production began in early 1877. It worked in the following way: an immersion liquid such as cedar oil, which had a distinctly higher refractive index than air, was placed between the specimen and the lens. This considerably improved the microscope's resolution, and offered many other benefits as well, such as fewer reflections. On 8 January 1879, Abbe was able to tell his friend Anton Dohrn about the major success that was the new product:

"Business at Zeiss has been very good recently. For 3 months now we have been working tirelessly to

fulfill the orders we have received. In particular, the new lenses (oil immersion) – and might I say it is a disgrace that you have yet to receive one of them because they are selling like hot cakes – have truly helped elevate the reputation of the Optical Works both in Germany and abroad. It would seem that over the past 6 months all institutes in Berlin with which we previously had no contact have placed orders for large microscopes."[8]

This outstanding success – microscope sales rose by 56 percent between 1878 and 1879 – also eradicated the retail business that had earned Zeiss no more than a modest additional income since the founding of his workshop. On 19 February 1880 Zeiss published the following announcement in the *Jenaische Zeitung*:

"I hereby respectfully announce that I have transferred my business in eyeglasses, telescopes, drawing apparatus, soldering blowpipes, thermometers,

Newspaper advertisement from 19 February 1880, "Jenaische Zeitung."

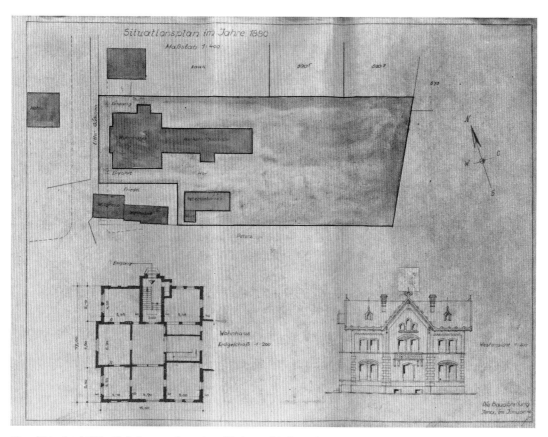

Map of Zeiss dated 1880 with the layout and west view of the house of Carl Zeiss (reconstructed to mark the 100th birthday of Carl Zeiss).

barometers, aerometers, etc. to Mr. Adolf Bleyer & Sohn of Johannisgasse and would request that you approach these gentlemen for all your needs in this regard."

In December 1879, Zeiss had already acquired a large plot of land outside the historical town center and planned to construct an edifice that would house both living quarters and a workshop as well as adjacent buildings. In a letter to the Jena municipal council, Zeiss explained how he intended to utilize the land:

"The ground floor of front building A will be reserved exclusively for the rooms needed to run my Optical Works (no manufacturing areas); the floor above it will serve as living quarters. The ground floor of annex B has been designated as an area for mechanical work, and the floor above it for optical work – and the work areas of Prof. Abbe and Dr. Riedel. The single-story adjacent building C is to be used as a carpentry workshop for my business and as a lavatory for the workers.
I hereby proclaim that my enterprise does not cause

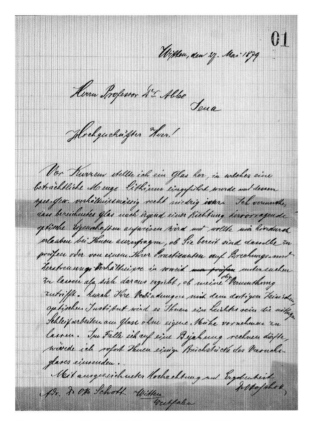

First letter from Otto Schott to Ernst Abbe, 27 May 1879.

ent on external suppliers for the glass he needed to produce the instruments in Jena – but the caliber of these suppliers was difficult to assess. This was by no means a new predicament, but Zeiss knew that setting up his own glass manufacturing facility was far too complex an endeavor without an expert on hand to make it all possible. But Zeiss had a stroke of good fortune: on 27 May 1879 Ernst Abbe received a letter from the young chemist Otto Schott (1851–1935), who had developed a new kind of glass in his parents' house in Witten, near Cologne. Could this be the solution to all of Zeiss's supply problems, for which his teacher Körner had searched in vain?

any disruption or botheration to my neighbors and that vehicular traffic does not reach levels exceeding that of an ordinary private residence."[9]

In the fall of 1880, so much progress had been made that Zeiss announced that the majority of his workers would be relocated on Monday 27 September. For the first time, production and development were housed in dedicated spaces. Yet the company was still missing something that would enable it to embark on a new era of growth: Zeiss was still depend-

1 Cf. Felix Auerbach: *Das Zeisswerk und die Carl-Zeiss-Stiftung in Jena.* Jena, 1925, p. 254.

2 Ernst Abbe: *Speech to commemorate the 50th anniversary of the optical workshop* in: *Gesammelte Abhandlungen vol. III.* Jena, 1906, p. 67.

3 Friedrich Merkel: *Das Mikroskop und seine Anwendung.* Leipzig, 1875, p. 52.

4 Carl Zeiss: *Description of Professor Abbe's Apertometer, with Instructions for Its Use,* in: *Journal of the Royal Microscopical Society 1* (1878), pp. 19–22.

5 ZEISS Archives, CZO-S 1.

6 Letter from Carl Zeiss to Georg Wagner dated 10 July 1859. Source: ZEISS Archives, CZO-S 16.

7 There is a contract dated 22 July 1876 which differs in terms of content but which was not signed. The actual partnership agreement dated 15 May 1875 was notarized on 25 August 1879, just before Roderich Zeiss officially became a partner in the company.

8 Cited by: Edith Hellmuth, Wolfgang Mühlfriedel: Zeiss 1846–1905. (Carl Zeiss. *Geschichte eines Unternehmens.* Bd. 1) Weimar, 1996, pp. 88–89. Source: ZEISS Archives BACZ 20387.

9 Letter from Carl Zeiss to the municipal council of the city of Jena dated 4 December 1879. Source: ZEISS Archives, BACZ 16155.

Chapter 5

Looking to the Future:
Preserving His Life's Work

« *Page 115: Carl Zeiss at the beginning of the 1880s.*

Carl Zeiss, Optische Werkstaette, Jena.
1846   1896.

Postcard to coincide with the anniversary in 1896; it shows the main site at the time and the three former workshops; sketched by Max Hunger, 1896.

Handing over the Baton:
Zeiss's Last Years (1880–1888)

Certificate for doctor of philosophy title (Dr. h. c.), dated 25 April 1880.

On 26 April 1880, with building work on the company's new premises in full swing, the Dean of Jena University's Faculty of Philosophy – well-known zoologist Ernst Haeckel (1834–1919) – wrote to his colleagues with a proposal to confer an honorary doctorate on Carl Zeiss. Haeckel argued that Zeiss's success as a lens designer and businessman made him the perfect candidate:

"As you are no doubt aware, the aforementioned Mr. Zeiss has spent 40 years engaged in the construction of simple and compound microscopes. By dint of a series of new improvements and ingenious inventions, he has succeeded in raising this vital instrument of observation to a remarkably high degree of accomplishment. As a result, no other optical works is currently capable of matching, let alone exceeding, the quality of Zeiss microscopes, thousands of which are now distributed across the globe. […] In light of the fact that the illustrious predecessors of Mr. Zeiss – the lensmakers Plössl, Fraunhofer and Hartnack – have received Doctor of Philosophy honoris causa degrees from various universities, it seems only fair that this honor should now also be bestowed on Mr. Zeiss by the very same university at which he once performed his studies of physics and optics. Since recent reports suggest that another university may be harboring the same intention, I have taken it upon myself to introduce this motion at the present time in order that our faculty may take precedence." [1]

It would appear that Haeckel's letter was merely a formality, because just two days later, on 28 April 1880, Carl Zeiss was officially awarded the title of Doctor of Philosophy *honoris causa*. It may equally be the case, however, that Haeckel's colleagues in Jena responded to his sense of urgency and wished to pre-empt any other university from bestowing the title on Zeiss. Whatever the reason, the son of a wood turner

The zoologist Ernst Haeckel.

from Weimar had now ascended to the ranks of the Grand Duchy's social elite.

The glass conundrum

Zeiss was now 63 years old, yet this recognition of his life's work by Jena University did not herald his retirement. There was still one essential piece missing from the Jena lensmaking industry: a factory dedicated to the production of optical glass. Ernst Abbe had been debating this issue with the chemist Otto Schott (1851–1935) since 1879, a scholarly exchange in which Zeiss initially played no active role. At that time, the art of producing glass for sophisticated technical applications was languishing at

Selecting and milling optical glass in the 1920s.

A young Otto Schott.

much the same level of development as microscope lensmaking had occupied before Abbe and Zeiss paved the way for mathematically precise lenses. The prestigious French and English makers of crown and flint glass on which Zeiss and his contemporaries depended made almost no distinction between producing glass for microscopes or for high-quality plate glass. Nobody understood how the optical qualities of different types of glass could be actively controlled by changing the composition of the melt. Yet the availability of glass with precisely defined parameters was clearly essential to the ongoing evolution of the optics industry. Abbe and Schott set to work on a step-by-step plan to make this a reality. Schott began with a series of melt experiments in his hometown of Witten. He sent the samples he produced to Jena where Abbe and his colleague

Paul Riedel analyzed them. Eventually Abbe decided they needed to work on a larger scale, so in January 1882 he set up a small glass technology laboratory in Jena.

From that point onward their quest took on a noticeably heightened intensity – and not simply because the laboratory's work was already yielding "very satisfactory results" in Abbe's words. Up to this point Germany had consistently failed to produce any manufacturers who could hold a candle to the French and English producers of optical glasses, so the experiments conducted in Jena also had a political dimension which was becoming increasingly evident. In November 1882, Schott was offered the opportunity to head up the optics department at a new Prussian State *Institute for the Advancement of Fine Mechanics* which was to be established in Berlin. The plan was for the institute to continue the glass technology experiments on a larger scale funded by taxes – in all likelihood with a stronger emphasis on national interests such as eliminating Germany's dependence on other countries for the production of military optics. At this point both Roderich and Carl Zeiss re-emerged to bolster Abbe's position. The three managing directors of the Jena Optical Works presented Schott with an alternative offer which he described in a letter to his friend Gottfried Brügelmann:

"They proposed that I work together with Abbe and Zeiss to build a factory in this very town to make standard optical glass and our special lenses. We would pool the investment capital required while

Scientific staff and lens designers, 1891. Pictured (back row, standing, from left to right): Albin Lautsch, A. Hartmann, E. Witte, Richard Schüttauf; (front row, seated, from left to right): Paul Riedel, Siegfried Czapski, Otto Schott, Paul Rudolph, Carl Pulfrich.

also requesting that the Prussian government support our firm and attempting to transfer interest in the Berlin undertaking onto us. It will probably come as little surprise to you that I did not ponder long on which side to take – you are, after all, very much aware of my sympathies for these local gentlemen."[3]

Schott heads to Jena

The plan was also warmly received in the German capital. The prospect of considerable private investment and the willingness of Zeiss and Abbe to devote their knowledge and infrastructure to the project

Letter from Otto Schott to his friend Gottfried Brügelmann, dated 3 December 1882.

were powerful arguments indeed. In early December 1882, Schott moved to Jena. Yet his assurance in the letter quoted above that they would "pool" the required capital does not appear to have been borne out by the facts. The funds to purchase a plot of land for Schott in Jena were actually mustered up by Zeiss, and the laboratory experiments conducted since 1882 had so far been financed by Abbe. Schott himself, much like Abbe in 1866, had no assets to speak of. It didn't take long for Schott to achieve some remarkable progress, however. He succeeded not only in controlling the properties of the optical glass, which was initially produced in small batches, but also in manufacturing relatively large samples with no impurities or internal stresses. Zeiss made the first microscope lens using Schott glass in the fall of 1883.

The results were extraordinary, offering a tantalizing glimpse of the remarkable improvements in optical instruments that could conceivably be achieved with the new material. This major step forward held out particular promise for telescopes with their correspondingly large lenses, a field in which poor glass quality had essentially created a bottleneck.

Yet it was still not clear whether the Prussian government would subsidize the Jena undertaking and, if so, to what extent. This uncertainty eventually placed Schott in such dire financial straits that he even considered abandoning his new home in the city on the river Saale altogether. In the fall of 1883, however, the three men finally received an assurance of state funding from Berlin. It is actually quite remark-

Jenaer Glaswerk Schott & Genossen in 1889.

able that the Kingdom of Prussia had the foresight at the time to invest in strategically important research projects outside of its territory. Soon after, Abbe, Schott, Roderich Zeiss and Carl Zeiss concluded an "agreement to establish a glass technology research institute" which came into force on 1 January 1884. The glass melting work began on 1 September 1884, and two weeks later Schott announced that production operations were now officially underway at the *Glastechnische Versuchsanstalt*.

Right from the start, the research institute was intended as a springboard for creating an industrial-scale glassworks. Its conversion into a general partnership under the name of "Jenaer Glaswerk Schott & Genossen" (Jena Glassworks of Schott and Associates) was provided for in the agreement, with the caveat that this should only be executed once they could

successfully compete with their foreign counterparts. The partners eventually took this step on 23 July 1885. The makers of precision optical instruments in Jena could now acquire most of the glass they needed from their affiliated company. The substantial capital costs incurred since the company was founded had largely been borne by the three owners of the Optical Works, namely Carl and Roderich Zeiss and Ernst Abbe, who together covered 61.5 percent of the investment. Schott's share came to a respectable 10.5 percent, while the remaining 28 percent was funded by state subsidies. Carl Zeiss's investment in the glassmaking company totaled 20,399 marks, only half as much as the sums put forward by Roderich Zeiss and Ernst Abbe. This provides some indication that the founder of the Optical Works was already scaling back his involvement in the business.

A new generation

As Zeiss senior gradually relinquished control of the tiller, his son Roderich increasingly came to the fore. Roderich had confidently assumed the role of commercial director, and Ernst Abbe would later state that Roderich played a key role in transforming the business from a small-scale operation into a modern factory:

"This man, Roderich Zeiss, who joined the company at the beginning of its fourth decade, gave it the fresh impetus it so badly needed to overcome those new challenges. And his connection to his father also aroused an entrepreneurial spirit which no longer shied away from the inevitable risks involved in converting any business into a large company. […] It was on his initiative that the company took the key organizational steps which were completed or at least initiated in this period, including the establishment of a proper system of business manage-

Roderich Zeiss, circa 1877.

ment, the acquisition of new sites with potential for expansion, the increased use of elemental forms of power and, above all, the beginnings of a more efficient division of labor in the manufacturing environment […]. Further steps included the decision to perform carpentry, casting and other support tasks in-house in order to free people's daily work from the many external encumbrances which previously stemmed from the need to depend on outside assistance." [4]

Abbe's criticism of Zeiss senior is mild yet unmistakable: it would seem that the founder of the Optical Works was finding it increasingly difficult to keep pace with the growth of his company and to shake off the cautious conservatism of a modest tradesman. Zeiss himself was aware that he could no longer meet the growing demands of the marketplace. In May 1883 he wrote the following lines to his children:

"Business is going so well that between 500 and 600 systems have been ordered, at least half of all our business now comes from England (perhaps including America), and no one can speak English apart from Abbe […] just as everyone seems to want the microscopes faster than ever […] and the two exhibitions in Amsterdam and Berlin took up so much of our time and effort […]. – And Löber is only completing an average of 150 systems a month. We are assailed by reproaches from Manchester today, from Cambridge tomorrow, and so on. It is a remarkably uncomfortable situation." [5]

Securing a business legacy

This huge surge in demand was primarily attributable to the unparalleled quality of the new apochromatic lenses. Gifted lensmakers had previously been

Jena and the Zeiss plant as seen from the south-west, 1891 (1: Carl Zeiss's house: administration on the ground floor, living quarters on the first floor, bookbinder's on the top floor; 2 and 3: building for mechanics and optics, 4: Abbe's house).

able to carve out a position in the market with their trial-and-error approach, but increasingly Zeiss's competitors were having to embrace scientifically based lensmaking as a matter of survival. Orders for Zeiss microscopes were coming in from academic and research institutions all around the globe, and inroads were steadily being made into the realms of physicians, hygiene specialists and material testers. The company was also sporadically starting to produce other optical products such as refractometers for measuring the concentration of solutions. But it was not until the 1890s that the new technological methods were applied to a greater variety of products including binoculars, camera lenses, astronomical devices, spectrometers and geodetic instruments,

opening up new areas of business which would further accelerate the company's growth.

In view of the steady rise in the company's sales and workforce, Carl Zeiss's decision to usher in a new generation was a sound one. The key foundations for this transition were laid in the summer of 1883 with revisions to the articles of association and a special supplement governing the relationship between Carl Zeiss and his son Roderich. This paved the way for Zeiss senior to depart the company – either of his own free will or in the event of his death.[6] The plan was for Roderich's share of the company's profits (originally 10 percent) to gradually increase at the expense of Carl Zeiss's share. The ultimate

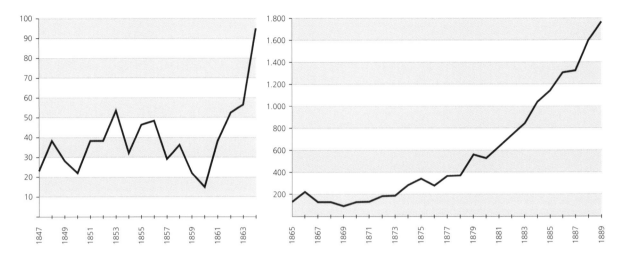

Number of microscopes sold each year: in 1847–1864 and in 1865–1889.

goal was to ensure an equal division of the profits between Carl Zeiss's son and Ernst Abbe by the time Zeiss passed away. The agreement stated that the other members of the Zeiss family would only receive corresponding payments for a short period of time but would ultimately be excluded from holding any share in the company. The primary goal of this provision was apparently to protect the business, though Carl Zeiss subsequently attempted to redress the balance in the following paragraphs of his last will and testament written in April 1886:

"In the following statement of how my estate will be apportioned among my heirs, I have taken into account the fact that my son Roderich was already placed in a highly favorable financial situation during my lifetime as a result of his inclusion in the business founded by myself and subsequently continued by myself and Professor Abbe. [...] The right contractually granted to him to receive a percentage of the share of the net profits generated by the company's primary business – a percentage that began at 25 % following my death but subsequently increased to 50 % – means that Roderich has been

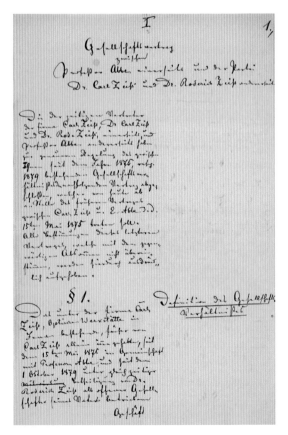

Partnership agreement between Ernst Abbe and Carl and Roderich Zeiss, 1883.

*endowed with a significant advantage over his oth-
er siblings which will continue to apply in the future.
He has, in a sense, inherited the intellectual capital
which I so painstakingly assembled in the course of
my life's work, and it is evident that this will provide
him with the source of a level of prosperity which
none of his other siblings can be expected to attain.*

*I hope that my son Roderich is himself acutely con-
scious of this significant advantage he enjoys over
his siblings. I am profoundly aware that his own
competence and prowess, which he has already
shown so readily, should suffice to ensure these
advantages come to fruition. And I am confident
that he will not consider it an affront if I treat my
remaining heirs slightly more favorably than him by
dint of some other provisions which render some
marginal advantages to them."* [7]

The collaboration between Roderich Zeiss and Ernst
Abbe did not always run smoothly, however. In
January 1885, for example, tensions ran high after
Roderich dismissed the young lens polisher Hertel
without notice due to his "trouble-making". [8] Abbe
protested, arguing that he should be consulted on
any decisions of this nature. Roderich, in turn, responded angrily, insisting on his right to make such
decisions independently and with the severity he
deemed necessary in order to maintain his authority.
He also announced that he intended to apply even
"greater energy" to ensuring "order" was maintained
in the company. These different styles of leadership
ultimately led to Roderich Zeiss leaving the company
in 1891. The dispute between these two figures

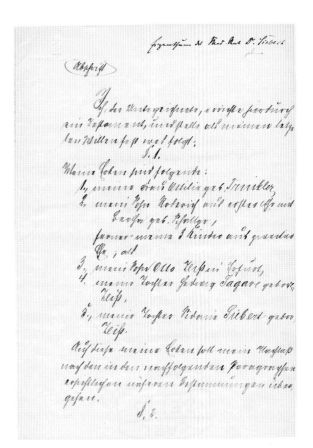

First page of the will and testament of Carl Zeiss.

is not our primary concern here, yet this episode
is pertinent to our biography of Carl Zeiss for one
very particular reason. The conflict with Abbe was
argued out in a letter which was obviously written
by Roderich, yet which also includes his father's
signature – in a somewhat shaky hand – in addition
to Roderich's own. In this context Zeiss senior hovers
in the background as a patriarch whose authority and reputation remain unbroken even though

The workshop, aka the "big hut" on Littergässchen, circa 1885.

his involvement in day-to-day business has virtually ceased. This conclusion is supported by the words of Abbe's colleague Siegfried Czapski, who joined the company in 1884 and wrote the following in his private correspondence one year later: "Zeiss senior no longer works at the company to any meaningful degree."[9]

Zeiss's final years

A number of letters have survived from the years 1884 to 1886 which bear testament to the long vacation trips taken by Zeiss senior in his later years. There is also some indication that he began cultivating roses in his old age. Reading also occupied much of his time, as testified by several biographers. The ZEISS Ar-

The Knight's Cross of the First Order presented to Carl Zeiss in 1886.

chives has preserved various orders which Zeiss sent to Jena bookseller Otto Deistung in 1886.[10] Works of popular history predominate – such as Friedrich Wilhelm Hackländer's tales of military exploits in *Krieg und Frieden* – as do works of light fiction such as *In der Dämmerstunde* by Pauline Schanz.

This gradual retreat into a somewhat bourgeois home life was unavoidable due to the steady deterioration in Zeiss's health. He suffered his first stroke in December 1885. Although he initially made a good recovery, there was no escaping the fact that Zeiss was getting

Carl and Ottilie Zeiss with daughter-in-law Franziska, née Thierbach, wife of Otto Zeiss, circa 1885.

old. In a letter written in 1886 to his son Roderich, Zeiss complains of persistent bouts of dizziness; and in November 1887 the newspaper *Jenaische Zeitung* reported that the father of Jena's optics industry had been "prevented by ill health" from attending the event at which his fellow partners and the workers at Schott's glass factory celebrated the 500th melt. In 1886, as he battled against the infirmities of old age, Zeiss was accorded a major honor by Carl Alexander, Grand Duke of Saxe-Weimar-Eisenach (1818–11901), who was fairly advanced in years himself. The sovereign presented his Court Mechanic with the *Knight's Cross, First Class, by Order of the White Falcon*, a distinction normally reserved for outstanding artists, professors, military personnel or officials. The fact that Zeiss, a businessman, was welcomed into this illustrious order speaks volumes for the elevated reputation which his field of optics and precision mechanics now enjoyed. One year later, Jena University's official mechanic was named as an honorary member of the *Society of Russian Physicians* at a congress held in Moscow. Zeiss was presented with a certificate which stated:

"The Congress unanimously declares that your new apochromatic lenses combined with the compensating eyepieces are the most accomplished ever produced in this field. The Congress firmly believes that this heralds a new era in the realm of microscopy research, setting us on the most promising of new paths." [11]

The founder nears the end of his voyage

Zeiss lived to see his company complete its 10,000th microscope, which led to much celebration on 24 September 1886. Yet his health was steadily deteriorating. He suffered further strokes and was soon confined to a wheelchair. Zeiss was now almost entirely absent from day-to-day business, though he

Letter of thanks to the Congress of Russian Doctors, 1887.

had not formally left the company. Therese Zeiss, who married Roderich in 1884, recalls her father-in-law's final months:

"Carl Zeiss's health deteriorated rapidly in the fall of 1888. [...] Withdrawn into himself, he now seemed thin and shrunken, with only his sparkling eyes to remind us of the upright, vivacious figure he once was with his full white beard and benevolent face. [...] He also continued to show an interest in our garden, in the flowers and fruits. A quiet mood had overtaken him, he seemed humble and unassuming, excessively so. His voice became softer and gentler, though his gaze was still as bright, alert and penetrating as ever. [...] Previously, when he celebrated a garden party with old friends, something he did frequently and gladly, then there was draft beer and grilled sausages aplenty. He would sometimes be very merry, though he was never loud. When all the children gathered around him, Carl Zeiss would enjoy telling jokes. These occasions always involved a selection of fine food and excellent wines in which his sons and sons-in-law were free to indulge." [12]

On Sunday, 2 December 1888, the weather forecast in the *Jenaische Zeitung* newspaper was for scattered clouds and temperatures of between two and nine degrees. Readers were also informed that the census conducted at the end of the previous month had counted 13,614 inhabitants in Jena, and that a meeting of the local trade association a few days earlier had seen locally registered companies present a number of novelties. The newspaper featured several advertisements by merchants looking to promote their products in the last few weeks before Christmas, a good example of which is that of Gehricke, a local mechanic and lensmaker advertising thermometers, spectacles and opera glasses. But for Carl Zeiss, now 72 years old, the bustle of life on the streets outside had slipped away. He was now dependent on full-time care from his wife Ottilie and nurses from the Sophienhaus Weimar charity. As evening approached on the first Sunday of Advent, his health deteriorated to such a point that his eldest son Roderich was summoned to his father's bedside. And

Ernst Abbe in 1876.

at half past midnight on 3 December 1888, Carl Zeiss passed away. On the Monday, six of his longest-serving employees carried his coffin to the Garnisonkirche church, nowadays known as the Friedenskirche. Many other workers had assembled along the route to pay their last respects to the company founder. The burial service took place two days later. Ernst Abbe broke off a trip to Switzerland and came straight from the railway station to the cemetery. The words of his eulogy were recorded for posterity – and after so many years working with Carl Zeiss he knew that his colleague's achievements extended far beyond his immediate vicinity:

"Standing here in front of the coffin of my departed friend, I am honor bound to bear testimony to the fact that this simple man, who strolled among us as the humblest citizen of our town, [...] that this man was one of the privileged few who found both the beginning and consummation of a new, fertile idea in his life's work and who is therefore destined to leave behind lasting traces of his existence. The fact is that everything before us today which emerged from

Anniversary plainsong penned by Carl Schäfer to mark the construction of 10,000 microscopes on 24 September 1886.

Manuscript (pages 1 and 3) of the eulogy given by Ernst Abbe upon the death of Carl Zeiss on 5 December 1888.

Carl Zeiss's efforts is no less than the living proof of an original notion which the deceased brought forth from himself and spent a quarter of a century striving to fulfill.

[...]

This notion of basing the practical construction of the microscope entirely on scientific theory and bringing all the art and craft of this process under its strict control – this is a notion which Carl Zeiss strove to implement regardless of the obstacles he faced, with a tenacity and perseverance which are only granted to those who firmly believe in the truth of the knowledge they have attained." [13]

Carl Zeiss was buried in the Johannisfriedhof cemetery next to the Garnisonkirche, his coffin adorned with a brass cross inlaid with objective lenses which had been produced at the Optical Works. The striking gravestone with its relief portrait of Zeiss in his final years still stands today. In remembrance of the deceased, the companies Zeiss and Schott issued a pension statute backdated to the day of Zeiss's death which "paved the way for such generous pension entitlements […] for employees […] that they could look to the future […] with an almost greater confidence than that of civil servants."[14] Two weeks after Carl Zeiss's death, Jena city council proposed that the streets bordering the company premises and the Zeiss family's home should be renamed as "Carl-Zeiss-Strasse" and "Carl-Zeiss-Platz" respectively. This proposal was approved unanimously by the council members.

Grave of Carl and Ottilie Zeiss in the Johannisfriedhof cemetery, circa 1900.

1 The original document of Haeckel's letter is not contained in Jena University's archives, but a facsimile has survived: *Zeiss-Werkzeitung Neue Folge* 8, issue 1, March 1933, pp. 7–8.

2 Letter from Ernst Abbe to Prof. Dr. Kleinenberg dated 9 November 1882. – Source: ZEISS Archives BACZ 10151.

3 Cited in: Jürgen Hendrich: *"Otto Schott"*, in: *Carl Zeiss und Ernst Abbe*, ed. Rüdiger Stolz, Joachim Wittig. Jena, 1993, pp. 267–268.

4 Ernst Abbe: "Gedächtnisrede zur Feier des 50jährigen Bestehens der Optischen Werkstätte", in: *Gesammelte Abhandlungen* ed. III. Hildesheim, 1989, pp. 80–81.

5 Letter from Carl Zeiss to his children dated 13 May 1883. Cited in: Edith Hellmuth, Wolfgang Mühlfriedel: *Zeiss 1846–1905*. Weimar, 1996, p. 149. – Source: Optical Museum Jena 12069.

6 Cf. Agreement made between Dr. Carl Zeiss and Dr. Roderich Zeiss in August 1883. – Source: ZEISS Archives CZO-S 428.

7 Last will and testament of Carl Zeiss dated 22 April 1886. – Source: ZEISS Archives CZO-S 458.

8 Cf. Letter from Carl and Roderich Zeiss to Ernst Abbe dated 30 January 1885. – Source: ZEISS Archives CZO-S 131.

9 Letter from Siegfried Czapski to a friend, fall 1885, cited in: Hellmuth/ Mühlfriedel: *Carl Zeiss 1846–1905*, p. 155. – Source: ZEISS Archives BACZ 22434.

10 ZEISS Archives CZO-S 6.

11 Cited in: Horst Alexander Willam: *Carl Zeiss*. Munich, 1967, pp. 131–132.

12 Cited in: Erich Zeiss: *Hof- und Universitätsmechanikus Dr. h. c. Carl Zeiss der Gründer der Optischen Werkstätte in Jena*. [No place of publication] 1966, p. 44.

13 *Nachruf auf Carl Zeiß. Gehalten von Prof. Ernst Abbe am Sarge in der Garnisonkirche in Jena 1888*, pp. 2–3. – Source: ZEISS Archives BACZ 11530.

14 Cited in: Axel Stelzner: "Carl Zeiß in der Jenaer Tagespresse," in: *Carl Zeiß und Ernst Abbe*, ed. Rüdiger Stolz, Joachim Wittig. Jena, 1993, p. 126.

The 20 Key Locations Where Zeiss Microscopes Were Sold Between 1847 and 1889

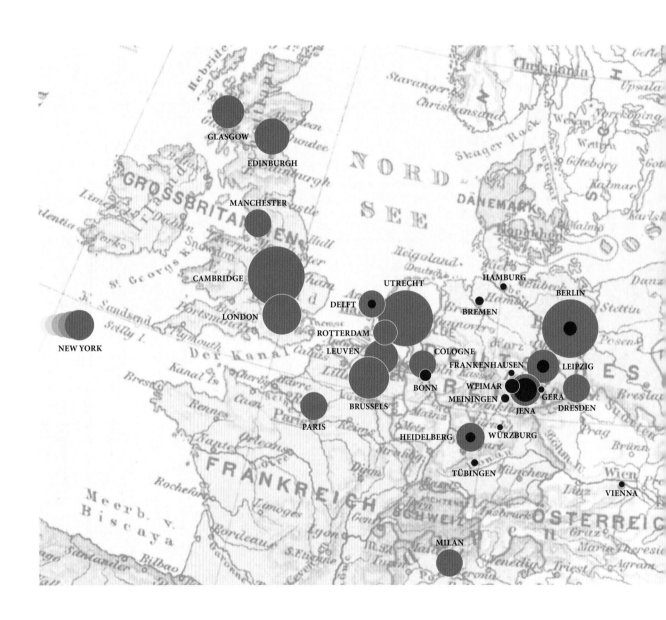

From 1847 onward, every microscope produced by Carl Zeiss was meticulously documented in the company's order books. An analysis of the records up to 1889, broken down by place and year of delivery, reveals that the company produced a total of nearly 15,000 microscopes over this period.

The diagram shows the 20 key locations where these microscopes were sold during the two periods of 1847 to 1869 and 1870 to 1889. The majority of orders in the first 20 years came from seats of royalty in the Jena region and some German university towns and major centers of commerce. Occasional sales were also made to areas which at that time formed part of Russia, in many cases facilitated by graduates of Jena University. The second period is dominated by major bastions of science and key centers of trade and commerce in Western Europe. New York also appears, and an increasing role was played by wholesalers in places such as Cambridge and Delft.

● 1847–1869   1,308 microscopes
● 1870–1889   13,228 microscopes

Source: ZEISS Archives BACZ 7710-7713. Compiled as part of a ZEISS Archives digital edition project led by Wolfgang Wimmer and funded by the Carl Zeiss Foundation. We would like to express our particular thanks to Maria Bischoff for validating the data and to the pupils and students who compiled the data.

How the Zeiss Workforce Evolved (1847–1889)

Number of employees, 1847–1879

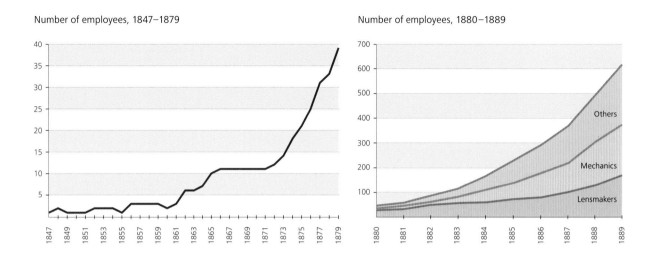

Number of employees, 1880–1889

Others

Mechanics

Lensmakers

Distance to employee birthplaces by occupational group

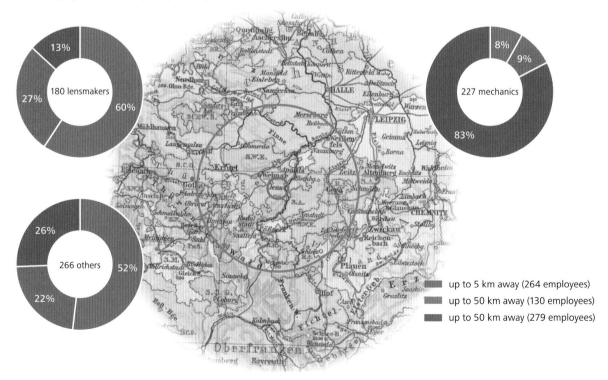

180 lensmakers — 13%, 27%, 60%

227 mechanics — 8%, 9%, 83%

266 others — 26%, 22%, 52%

- up to 5 km away (264 employees)
- up to 50 km away (130 employees)
- up to 50 km away (279 employees)

Carl Zeiss began hiring employees in increasing numbers from the early 1860s onwards as the sales of compound microscopes gathered pace. This process stalled somewhat in the mid-1860s when Zeiss began working with Ernst Abbe, but then picked up again in the early 1870s. From 1881 onwards the company grew at a tremendous rate, fueled by a surge in orders and the rapid expansion of the new works on the site of what would subsequently become the main Zeiss factory.

The company employed mechanics from all the German-speaking regions, though most of the new lensmakers came from the area around Jena. Over half the lensmakers and clerks spent the rest of their working lives at Carl Zeiss. The company was keen to hold onto them because they typically had Zeiss-specific expertise and know-how. The situation was very different when it came to mechanics, however, only around one fifth of whom pursued a long-term career within the company. The category 'Miscellaneous' covers multiple professions including joiners, bookbinders, metalworkers, foundry workers and others. It also includes white-collar clerks and university graduates, the number of which increased steadily from 1876 onwards, essentially distancing the company from its traditional workshop roots.

Unskilled workers were primarily hired between 1886 and 1888. This category also included many of the female employees known as 'working girls,' most of whom left the company soon after they had joined it. In fact of the 33 girls originally employed, 22 no longer worked for the company by 1890. Otherwise women were few and far between in the ranks of Zeiss employees. Seven adult women had been employed until 1889, including a saleswoman in the retail outlet which still existed at that time, three female workers, two female packers and a lady performing optical ruling, who was most likely a highly qualified

specialist in her field. Women did not enter the lensmaking business to any significant degree until 1900.

Many of the male unskilled workers left the company soon after they were hired, in some cases after just a few days. Others are subsequently listed as skilled workers, suggesting they had undergone some vocational training. The company began a systematic training program in 1884, though this was initially devoted almost entirely to lensmakers.

Length of service by occupational group

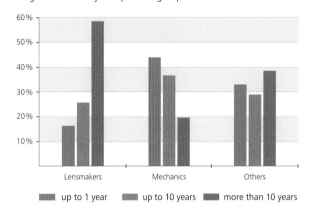

Hiring of young workers and apprentices

Sources: ZEISS Archives BACZ 1737–1738, 23712–23714.

"A Role Model for Future Generations of Entrepreneurs":
An Interview with Dr. Dieter Kurz, Chairman
of the Foundation Council of the Carl Zeiss Foundation

Can you tell us about the first time you encountered Carl Zeiss as a historical figure?
That would be shortly after I joined Carl Zeiss in Oberkochen in 1979. The main administration building housed an assembly room, the Ernst Abbe room, where the busts of Carl Zeiss and Ernst Abbe were on display. That's when I got an idea of what our company founder actually looked like. Quite frankly, that was more or less the long and short of it: while I was already familiar with Abbe as a physicist from my time at university and knew about the decisive role he played for the culture of remembrance at the two Carl Zeiss companies in East and West Germany, as a young employee I was more or less in the dark when it came to Carl Zeiss. This also reflects what was

considered historically important: the period that followed Abbe joining the company is well-documented. However, I don't think all that much was written about the 20 years in which Carl Zeiss single-handedly built his company from scratch.

That is perhaps a consequence of the first historical accounts written by people who subscribed to the teachings to Abbe, such as Friedrich Schomerus and Moritz von Rohr.
The fact nevertheless remains that the company wouldn't be what it is today without Carl Zeiss. For me, there are two decisive factors that paved the way for the company's future: first, Carl Zeiss turned his attentions to developing and manufacturing

microscopes very early on even though this was an area in which he faced stiff competition. Second, he recognized the importance of having a team of qualified staff working in precision mechanics and optical manufacturing who had the necessary tools at hand and were truly dedicated to the company. Ernst Abbe was able to build on Zeiss's core beliefs to advance his own ideas for social reform. There's one more thing I deem important: Ernst Abbe and Carl Zeiss were able to convince Otto Schott to help them set up the *Glastechnisches Laboratorium* in Jena – a great victory if you ask me. Today you'd say something like: Zeiss and Abbe invested a substantial amount of money in Schott's start-up to expand their own value chain. This step alone is evidence that Carl Zeiss was a bona fide entrepreneur and far more than just a master craftsman who happened to win over the right people.

The partnership between Carl Zeiss, Ernst Abbe and Otto Schott resulted in a company that thrived on the interactions between industry and science. To what extent was this still characteristic of ZEISS when you became President and CEO in 2001?
Many illustrious scientists still approach ZEISS directly to help them advance the development of new equipment for their research. For many customers, in fact, ZEISS is the company that can realize highly complex and high-quality solutions. For me, this brand image is just one example of the impact the legacy of Carl Zeiss and Ernst Abbe has on us today. Our ownership structure is also significant in this respect. A company in the hands of a foundation enjoys more respect in

the public arena as it shows that, alongside its legitimate economic interests, it also champions social responsibility. This structure also equipped the company with the tools it required to surmount difficult periods in German history.

The foundation was established as a result of the efforts made by Ernst Abbe. Do you see any historical link whatsoever between Carl Zeiss and the foundation that bears his name?
What we can say for certain is that Abbe consciously chose the name for the Carl Zeiss Foundation. What we don't know is whether there was any debate between the two men regarding the transformation into a foundation company. But Abbe certainly believed that the move was in keeping with his good friend's interests.

You touched on the fact that ZEISS still enjoys an excellent reputation among scientists today.
Indeed it does. Our focus on leading research is one of our defining features – and there are certain challenges that come with that: as an industrial company, ZEISS relies on the fact that technical developments do not end with prototypes or are only produced in small batches but that they are transformed into economically viable products. There's often a gap between what leading scientists themselves want to achieve in their work and the expectations that most users from the same discipline have of a product. This is a dilemma I was often faced with when I was in charge of electron microscope sales for ZEISS in the USA at the end of the 1980s. Some of the functions offered by ZEISS were developed to meet the needs

of specific key customers, while other potential buyers saw this as nothing more than generating unnecessary costs, which was hard to defend in price negotiations. Innovation must also make good business sense. This is where we can take a leaf out of the book of our founding fathers – after all, the collaboration between Carl Zeiss and Ernst Abbe to design microscope optics not only served to enhance precision but also reduced the amount of wastage, thus ensuring better prices and higher profit margins.

The transition from the traditional trial and error process to production based on scientific calculations was revolutionary. How would you compare it to the optical industry today? Does progress happen gradually or are we on the cusp of a new revolution?

That all depends on your perspective. At ZEISS, for example, you could draw a continuous line from the first microscope to the latest leading innovations, such as lithography lenses. The latter is based on the principles established by Carl Zeiss in the early decades that followed the company's founding: tailored solutions, ultimate precision, outstanding optical materials and exact calculations for all optical properties prior to production. Abbe's limit, the resolution limit for classic light microscopes, still plays a major part in the development of optics for microchip production. Even so, the optical industry will have to prepare itself for a change of similar proportions to that which accompanied the scientific turning point for microscope production in the 19th century. The ever greater available processing power and the advancements in digital image processing mean that innovations today are much less focused on enhancing the optical and mechanical aspects of a device than on the intelligent processing of image data. It is thus becoming increasingly important for ZEISS to be au fait with the latest software. If Carl Zeiss and Ernst Abbe now had to mull over how they could make visible that which was previously invisible, my guess is they wouldn't just look for a skilled glass maker like Otto Schott, they would also be on the lookout for IT specialists and software engineers.

So digitization is forcing ZEISS to develop skills that go beyond the realm of classic business activities. Isn't this similar to how Carl Zeiss himself made the transition from mechanics to optics?

What is striking about Carl Zeiss is that not only was he flexible enough to call into question prevailing practices in his field, he also encouraged others to step outside of their comfort zone. A classic example of this is August Löber, who was hired as Zeiss's first apprentice in 1847 at the age of 17 and who went on to become Head of Optical Production. We should also bear in mind that Ernst Abbe only started delving into the basics of optics in microscopy once Zeiss piqued his interest. That's why I see Carl Zeiss as an excellent motivator and pacesetter who also had a keen sense for scouting out talented individuals and recognizing technological trends. If we see the founder of ZEISS in this way, then he is an excellent role model for current and future generations of entrepreneurs.

....

Dr. Dieter Kurz (*1948 in Tübingen) studied physics at the University of Tübingen where, as a postgraduate student, he also earned the title Dr. rer. nat. He joined ZEISS in 1979, became a Member of the Executive Board in 1999 and served as President and CEO from 2001 to 2010. Since March 2012 he has been Chairman of the Foundation Council of the Carl Zeiss Foundation and Chairman of the Supervisory Board of both Carl Zeiss AG and SCHOTT AG.

Key Dates in the Life of Carl Zeiss

Youth and education

11 September 1816	Born in Weimar as the fifth of twelve children of Johanna Antoinette Friederike (1786–1856), née Schmith, and the art wood turner Johann Gottfried August Zeiss (1785–1849)
1832	Graduated from grammar school
1834–1838	Apprenticeship under university mechanic Dr. Friedrich Körner (1778–1847) in Jena while studying mathematics
1838–1845	Journeyman years: travels to Stuttgart, Darmstadt, Vienna (Rollé & Schwilgué, lectures at the Imperial and Royal Polytechnic Institute) and Berlin
1845–1846	Period spent in Jena "to primarily study chemistry and advanced mathematics" and acquire citizenship of and the right of abode in Jena

Establishment of the mechanical workshop in Jena

17 November 1846	Opening of a workshop ("atelier for mechanics") on Jena's Neugasse
1 July 1847	Relocation to a larger workshop on Wagnergasse
1847	The first simple microscope is produced
29 May 1849	Marries Bertha, née Schatter (1827–1850)
23 February 1850	Birth of son Roderich (1850–1919); Bertha dies in childbirth
17 May 1853	Marries Ottilie, née Trinkler (1819–1897)
25 February 1854	Birth of son Otto (1854–1925)
27 September 1856	Birth of daughter Hedwig (1856–1935)
1857	The first compound microscope is produced
1 May 1858	Relocation to a workshop on Johannisplatz
12 July 1858	Becomes Deputy Master of Weights and Measures in Jena
10 September 1860	Appointed university mechanic
23 July 1861	First honorary prize and gold medal at the 2nd General Thuringian Trade Exhibition
1 October 1861	Birth of daughter Sidonie (1861–1920)
1863	Zeiss becomes court mechanic
28 May 1866	Production of the 1,000th microscope

Collaboration with Ernst Abbe

3 July 1866	Dr. Ernst Abbe is brought in to assist Zeiss as an independent researcher (in 1870 also a professor at the University of Jena) From 1872 onwards, all ZEISS microscope optics are built in line with Abbe's calculations
1 January 1875	Creation of Carl Zeiss health insurance
15 May 1875	Ernst Abbe becomes silent partner
1876	Son Dr. Roderich Zeiss joins the company
1877	First microscope lens with homogeneous immersion produced
1878	Daughter Hedwig marries the grammar school teacher Konrad Sagawe (1858–1935)
1880	Carl Zeiss receives honorary doctorate (Dr. phil. h. c.) from the University of Jena
1880	Construction of a private residence on Littergässchen, today Carl-Zeiss-Strasse, and first factory building on what will later be the site of the main factory

A growing family

1881	Son Otto marries Franziska Thierbach (1862–1885)
1882	Daughter Sidonie marries the physician Johann Lucas Siebert (1841–1913)
26 January 1882	Birth of granddaughter Charlotte (1882–1945)
20 May 1882	Birth of grandson Franz (1882–1882)
1883	Son Roderich marries Therese Schatter (1864–1949)
2 July 1883	Birth of granddaughter Johanna (1883–1943)
6 July 1883	Birth of grandson Berthold (1883–1945)
1 January 1884	Founding of the *Glastechnisches Laboratorium* (glass laboratory), later to become *Jenaer Glaswerk Schott & Gen.* (today SCHOTT AG) by Otto Schott, Ernst Abbe, Carl Zeiss and Roderich Zeiss
16 March 1884	Birth of granddaughter Elisabeth (1884–1967)

Old age and death

December 1885	Carl Zeiss suffers a minor stroke
24 September 1886	10,000th Zeiss microscopes produced
19 November 1886	The Grand Duke of Saxe-Weimar-Eisenach names Carl Zeiss the Knight's Cross, First Class, by Order of the White Falcon
25 November 1888	Birth of granddaughter Elisabeth (1888–1958)
3 December 1888	Carl Zeiss dies in Jena

Notes on Sources

In recent years, the records held at the ZEISS Archives on the company founder and the period in which he lived were edited on numerous occasions: all of Zeiss's letters which are available in the corporate archives were transcribed in full. Microscope and employee lists were digitized and statistically analyzed. The bookkeeping records kept by Carl Zeiss from 1848 to 1863, the 'Manual,' are now only available as a scan. The sources named are an indispensable part of this book. This material will soon be published in digital form (www.urmel-dl.de/).

Added to that are several reference works only accessible as manuscripts, such as the one compiled in 1957 by Herbert Koch entitled *Unbekanntes über Leben und Wirken von Carl Zeiss*. Other archives, for example the Main State Archives in Weimar and the Friedrich Schiller University Archives in Jena, contain limited historical information on Carl Zeiss. In spite of this, we were able to gain key additional insights by accessing these archives. The locations of little-used documents are specified in the comments. Overall, however, our account is based primarily on the existing literature; due to its temporal proximity to historical events, it is more of a source than a scientifically accurate account. Below is a selection of key reference works.

Selected Literature

- Abbe, Ernst: "Gedächtnisrede zur Feier des 50jährigen Bestehens der Optischen Werkstätte (1896)," in: *Gesammelte Abhandlungen III. Vorträge, Reden und Schriften sozial-politischen und verwandten Inhalts*. Jena, 1906 [reprint: Hildesheim, 1989], pp. 60–101.
- Auerbach, Felix: *The Zeiss Works and the Carl Zeiss foundation in Jena. Their scientific, technical and sociological development and importance popularly described* 2nd edition, Jena 1904.
- Esche, Paul Gerhard: *Carl Zeiss. Leben und Werk*, ed. Jena City museum. (Works from the Jena City museum, no. 4), 2nd edition, Pössneck, 1977.
- Gerlach, Dieter: "Carl Zeiss (1816–1888)," in: *Mikrokosmos 77 issue 9* (1988), pp. 263–273.
- Gerlach, Dieter: *Geschichte der Mikroskopie*. Frankfurt am Main, 2009.
- Hellmuth, Edith; Wolfgang Mühlfriedel: *Zeiss 1846–1905. Vom Atelier für Mechanik zum führenden Unternehmen des optischen Gerätebaus. (Carl Zeiss. Die Geschichte des Unternehmens*, ed. Wolfgang Mühlfriedel, Rolf Walter, vol. 1) Weimar, 1996.
- Gubas, Lawrence J.: *A Survey of Zeiss Microscopes 1846-1945*. Las Vegas, Nevada, 2008.
- Gubas, Lawrence J.: *Zeiss and Photography*. Las Vegas, Nevada, 2015.
- Meinl, Hans: "The history of the Optisches Museum in Jena – Part I," in: Treasury of Optics. *The Collections of the Optisches Museum Jena*. Jena, 2014, pp. 15–38.
- Günther, Norbert; Bower, David I. [Transl.]: *Ernst Abbe – Creator of the Zeiss Foundation*. 2016
- Harold Moe: *The Story of the Microscope*. Rhodos Int. Science and Art Publishers, Denmark, 2004.
- Paetrow, Stephan: *Birds of a Feather. 20 Years of Reunification at Carl Zeiss*. Hanseatische Merkur, Hamburg, 2011.
- Rohr, Moritz von: *Zur Geschichte der Zeissischen Werkstätte bis zum Tode Ernst Abbes*. Jena, 1936.
- Schomerus, Friedrich: *Geschichte des Jenaer Zeisswerkes 1846–1946*. Stuttgart, 1952.
- Willam, Horst Alexander: *Carl Zeiss – Mensch und Werk* ("Zum 75-jährigen Todestag am 3. Dezember 1963," in: *Tradition. Zeitschrift für Firmengeschichte und Unternehmerbiographie*). Vol. 9, issue 2, 1964, pp. 58–69.
- Willam, Horst Alexander: *Carl Zeiss. 1816–1888*. (Sixth supplement of *Tradition. Zeitschrift für Firmengeschichte und Unternehmerbiographie*) Munich, 1967.
- Zeiss, Erich; Friedrich Zeis: *Hof- und Universitätsmechanikus Dr. h. c. Carl Zeiss, der Gründer der Optischen Werkstätte zu Jena. Eine biographische Studie aus Sicht seiner Zeit und seiner Verwandtschaft*. [No place of publication] 1966.

Photography

Weimar City Museum (pp. 8, 10 top, 11, 13, 14 left),
Wikimedia Commons / public domain (pp. 9, 12, 29, map on
pp. 130–131),
Jan-Christoph Hauschild (p. 14 right),
Stiftung Stadtmuseum Berlin (p. 31),
Vienna Museum (p. 32),
Private archives of Volker Loeschcke, Aarhus, Denmark (p. 35),
Jena University Archives, collection V, dept. LXXXIV, photo
folder (p. 42),
Jena Botanical Institute, former garden records (p. 45 right),
Bavarian Army Museum, Ingolstadt (p. 53),
Fraunhofer Institut (p. 60),
Jena City Museums (pp. 70 bottom, 74 bottom),
Jena University Archives, photo collection (p. 71 bottom),
Timo Mappes, www.musoptin.com (pp. 80, 93 top, 109 top),
Library of Congress Prints and Photographs Division Washington, LC-DIG-pga-00497 (p. 88),
Humboldt University Archives, Berlin (p. 91 right),
Rathenow Cultural Center (p. 106 right),
Dieter Herfurth (p. 125 top),
GEI-Digital + http://gei-digital.gei.de/viewer/resolver
?urn=urn:nbn:de:0220-gd-9923482 (map on p. 132)

The photographs on pages 16, 18, 64, 66, 96, 98 and 142 were
taken by Anna Schroll and those on pages 134 and 136 by
Albert Kunzer.

All other images were taken from the ZEISS Archives.

Acknowledgments

The editorial team, March 2016 (from left to right): Stephan Paetrow, Bernd Adam, Dr. Wolfgang Wimmer, Marte Schwabe.

Aside from the core project team, a large number of people also spared no effort in making this book possible. We would like to thank Eric Betzig, Michael Kaschke, Dieter Kurz and Kathrin Siebert, all of whom provided us with fresh perspectives on the man that was Carl Zeiss in the interviews they gave. Thank you to Albert Kunzer and Anna Schroll for their photography during the interviews. We would also like to thank Timo Mappes, who did a great job of interviewing Eric Betzig and assisted us with questions relating to the history of microscopy. This book contains numerous photographs from Timo Mappes's extensive collection of historical microscopes (www.musoptin.de). Dieter Brocksch, Michael Kaschke, Kathrin Siebert and Gudrun Vogel were also instrumental in providing us with comments on the manuscript.

A range of institutions granted us access to documents and image material, and we were able to benefit from the excellent advice and research skills of the staff there time and again. Special thanks goes to the Thuringia State and University Library and the Hausknecht herbarium (Hermann Manitz), the Jena City Archives (Constanze Mann) and the Weimar City Archives (Jens Riederer), the State Archives in Berlin (Martin Luchterhand), the Hessian State Archives in Darmstadt and the Main State Archives in Weimar (Bernhard Post, Dagmar Blaha, Frank Boblenz), the Friedrich Schiller University Archives in Jena (Joachim Bauer, Margit Hartleb) and the Humboldt University of Berlin. The city museums in Jena (Birgitt Hellmann), Weimar (Marina Reichardt), Berlin and Vienna and the Cultural Center in Rathenow

(Bettina Götze) lent us their support with their substantial image collections. Ms. Ulrike Müller-Harang from the Weimar Classics Foundation carried out in-depth research into the life and work of August Zeiss. Dietrich Herfurth assisted us with his knowledge on Orders. Wolfgang Albrecht afforded us a glimpse into his studies related to his ancestor Friedrich Körner.

A key source for this biography were the digital versions of the letters and documents penned by Carl Zeiss, Ernst Abbe and Otto Schott themselves. Thanks to Alexandra Stelzig and Johannes Weiss, who transcribed many of the handwritten documents, thus making them much easier for us to work with. We would like to thank the Carl Zeiss Foundation (Klaus Herberger) for financing this project. The collected statistics on microscopes and employees held at the ZEISS Archives and based on the original sources were also extremely helpful. We would also like to thank the following trainees for their work on this extensive digitization project: Timo Bachmann, Laurenz Paul Bahr, Veruschka-Meike Jähnert, Amin Laroussi, Markus Lindner, Selina Schottmann, Emilienne Thieme, Laura Thomann and Tan Tran, to name but a few.

Employees from the ZEISS Archives also contributed to the project: Maria Bischoff, Stefan Lux, Paul Mokry, Dominique Schmied, Alexandra Stelzig and Clemens Uhlig. We would also like to thank Karl-Heinz Wilke for his work on our virtual museum (www.zeiss.de/archiv). Marte Schwabe was in charge of image research for the biography and was the glue that held the entire project together; without her, none of this would have been possible. At *timefab*, the historical research agency involved in this project, Tim Sander proofread and provided a critical assessment of the majority of the texts before passing them on to the editors. Dorothee Rheker-Wunsch worked on the project at the Böhlau publishing house. Clive Poole, James Humphreys, Marina Stephanou and Charles Taggart translated the German text. Constanze Lehmann assumed responsibility for the final proofread. Last but not least, we would like to thank Bernd Adam for his unwavering professionalism and creativity regarding the layout and graphic design.

Jena and Leipzig, March 2016

Wolfgang Wimmer   Stephan Paetrow

Dr. Carl Zeiß.